Sustainable Rice Production - Challenges, Strategies and Opportunities

Edited by Min Huang

Published in London, United Kingdom

Sustainable Rice Production - Challenges, Strategies and Opportunities
http://dx.doi.org/10.5772/intechopen.104284
Edited by Min Huang

Contributors
Banoth Madhu, Bhimireddy Sukrutha, Nunavath Umil Singh, Govada Venkateswarao, Jannatul Ferdous, Mohammad Abdullah al Asif, Farah Mahjabin, Israt Jahan Riza, Mohammad Mofizur Rahman Jahangir, Hosahalli Buthanna Narasimhamurthy, Mukesh Sehgal, R. Ganesha Naik, Shaw Watanabe, Olayemi Mikail Mikail Olaniyi, Abraham Ayegba Alfa, John Kolo Alhassan, Morufu Olalere

First published in London, United Kingdom, 2023 by IntechOpen
IntechOpen is the global imprint of INTECHOPEN LIMITED, registered in England and Wales, registration number: 11086078, 5 Princes Gate Court, London, SW7 2QJ, United Kingdom

British Library Cataloguing-in-Publication Data
A catalogue record for this book is available from the British Library

Additional hard and PDF copies can be obtained from orders@intechopen.com

Sustainable Rice Production - Challenges, Strategies and Opportunities
Edited by Min Huang
p. cm.
Print ISBN 978-1-83768-101-3
Online ISBN 978-1-83768-102-0
eBook (PDF) ISBN 978-1-83768-103-7

Meet the editor

Dr. Min Huang received his bachelor's degree in Biological Science from Zhejiang Normal University, China, in 2005. He obtained a master's degree in Crop Cultivation from Guangxi University (GXU), China, and a Ph.D. in Crop Cultivation and Farming System from Hunan Agricultural University (HNAU), China, in 2008 and 2011, respectively. He successively held the positions of assistant and associate professor in the Department of Agronomy, GXU, from 2012 to 2014. He joined the faculty in the Department of Agronomy at HNAU as an associate professor in 2015 and was promoted to professor in 2017. Dr. Huang served as a visiting fellow at the International Programs-College of Agriculture and Life Sciences, Cornell University, USA, in 2017 and 2018.

Contents

Preface

The global population will be nearly 10 billion by 2050, which means there will be about 2 billion more mouths to feed than there were in 2022. Rice is one of the most important staple foods in the world, on which more than half of the population relies for more than 20% of their daily caloric intake. It is projected that global rice production will need to increase by 70% by 2050 to meet the food demands of the world's growing population.

Satisfying future rice demands will mainly depend on our ability to improve rice productivity rather than on the area enlargement of rice paddies because of space limitations caused by urban expansion. At the same time, we also require efforts to improve rice quality because the demand and consumption of high-quality rice will increase as living standards improve. Also importantly, we need to develop new technologies and strategies to overcome the constraints (e.g., climate change, soil degradation, and biodiversity loss) confronting the sustainable development of rice production.

This book describes some challenges, strategies, and opportunities for sustainable rice production. Chapter 1 introduces the distribution, symptoms, biology, survival strategies, and control measures of rice root-knot nematode. Chapter 2 elaborates on the responses of plant morphological, physiological, and biochemical traits to drought stress in rice as well as the conventional breeding approaches and molecular basis for improving drought tolerance in rice. Chapter 3 introduces principles and management practices of gaseous losses of nitrogen from rice fields. Chapter 4 presents a process and architecture for smart rice precision farming schemes in Sub-Saharan Africa. Finally, Chapter 5 introduces the potential health benefits of brown rice and recently developed low-protein fermented brown rice.

I am grateful to Dr. Haiming Tang at the Hunan Soil Research Institute, Dr. Xiaowu Pan at the Hunan Rice Research Institute, and Dr. Zhong Ding at the Hunan Agricultural University for their help with reviewing chapters.

Dr. Min Huang
Rice and Product Ecophysiology,
Key Laboratory of Ministry of Education
for Crop Physiology and Molecular Biology,
Hunan Agricultural University,
Changsha, China

Chapter 1

Rice Root-Knot Nematode (*Meloidogyne graminicola*): A Major Menace in Rice Production

Hosahalli Buthanna Narasimhamurthy, Mukesh Sehgal and R. Ganesha Naik

Abstract

Rice is an important major staple food crop of the world which is affected by various biotic and abiotic stresses. Among biotic stresses, plant parasitic nematodes are considered as major constraints. However, of late, *Meloidogyne graminicola* has emerged as pest of International importance and it is considered as number one enemy of rice crop. Being a soil borne and hidden organism in rice causes a yield loss up to 80%. Due to its adaptation, distribution, broad host range and ability to survive under different abiotic factors especially physical and chemical properties of soil, such as soil pH, organic carbon, EC, nutrition, temperature, soil type, moisture, etc., The management is a challenging issue due to non-availability of nematicides and also effective management practices all these factors represents, *Meloidogyne graminicola* a serious menace for rice production. Considering the impact of this nematode in rice production a literature is mainly focusing on distribution, symptoms, biology, survival strategies and management practices.

Keywords: *Meloidogyne graminicola*, rice root-knot nematode, plant parasitic nematodes, management, rice production

1. Introduction

Rice (*Oryza sativa* L.) is the important staple food for more than half of the world's population [1] and depends on rice for more than 20% of their daily calorie intake [2]. More than 90% of the world's rice area is in Asia, which is the home for more than half of the world's poor, and more than half of the world's rice cultivators [3]. Rice is affected by various biotic and abiotic stresses, among biotic stresses the diseases caused by fungi, bacteria, virus and nematodes are considered as major threat. They affect the increasing productivity of rice. Among various diseases, blast, bacterial blight, sheath blight, sheath rot, brown leaf spot, false smut, rice tungro virus and rice root-knot nematode are economically important and cause significant yield loss [4]. Recently, rice root-knot nematode (*Meloidogyne graminicola*) as becoming serious threat in both upland and low land rice [5].

Root-knot nematode (*Meloidogyne spp.*) is biotrophic, sedentary endoparasite and perfect examples of highly adapted and evolved root parasitism. They are one of the major menace in the production of field and horticultural crops throughout globe. *Meloidogyne sp.* create permanent feeding sites to enjoy continuous supply of nutrient and water from the infected host and continuous feeding results in the production of galls or knots [6]. The resultant infection in host is due to the hypertrophy of vascular tissues and hyperplasia of root cortex cells [7]. Besides the galls on roots, an infected plant also exhibit poor growth, stunted growth, reduced tillers, unthriftiness and general wilt symptoms the damage is aggravated by the parasites interaction with other pathogens such as fungi and bacteria [8]. In the present era of advanced farming system, rice is prone to attack by various of abiotic and biotic stresses which results in reduce the crop yield, which includes tiny hidden organisms *i.e.,* plant-parasitic nematodes [9, 10]. Plant parasitic nematodes (PPNs) pose a major threat to the rice is also attacked by a wide array of nematodes [7, 11]. Over 200 species of plant parasitic nematodes have been reported to be associated with rice [12] and are becoming increasingly important in the rapidly changing production system of rice. Rice root-knot nematode, *Meloidogyne graminicola* Golden and Birchfield 1965 has emerged as a pest of international importance [13]. Among the major PPN species attacking rice, the root-knot nematode (RKN), *Meloidogyne graminicola* Golden and Birchfield, 1965 is considered as a major threat to rice alone has been reported to cause 50–75% yield loss under different conditions [14, 15]. For example in countries like China it has reported 85% [16], in Bangladesh under lowland rain fed rice losses can range between 16 and 20%, while in India, 16 and 32% under irrigated conditions and between 11 and 73% under flooded conditions [17, 18]. Since, its short life cycle, wider adoptability and wide host range, including many weed hosts that are commonly found in rice fields, make this species difficult to manage [13, 19].

2. Distribution of *M. graminicola*

2.1 Global distribution

Meloidogyne sp. a destructive nematode of rice it is widely distributed in various rice growing areas of the world; it was first described in 1965 from grasses and oats in Louisiana as *M. graminicola* [20]. *M. graminicola* distributed in different parts of the globe *viz.,* Bangladesh, Philippines, China, Nepal, India, South East Asia, Burma, Laos, Thailand, Vietnam and USA. However, *M. oryzae* in Surinam of irrigated rice, *Meloidogyne graminicola* in Costa Rica, Cuba, Egypt, Ivory Coast, Nigeria, South Africa and Japan, *M. javanica* in Brazil, Egypt, Comoro Islands, Nigeria and Ivory Coast, *M. arenaria* in Nigeria, Egypt and South Africa and *M. salasi* in Costa Rica and Panama on upland rice [21]. Pakistan [22]. It is also found in the United States and Latin America, and was recently reported in Africa and Europe [23–27] (**Figure 1**).

2.2 In India

Since, *Meloidogyne graminicola* was observed for the first time during 1969 in association with rice [28] in India, its prevalence has been recorded from all the rice growing states of the country namely, Assam, Andhra Pradesh, Andaman and Nicobar Islands, Bihar, Delhi, Himachal Pradesh, Haryana, Jammu and Kashmir, Karnataka, Kerala, Madhya Pradesh, Odisha, Punjab, Telangana, Tripura, Tamil

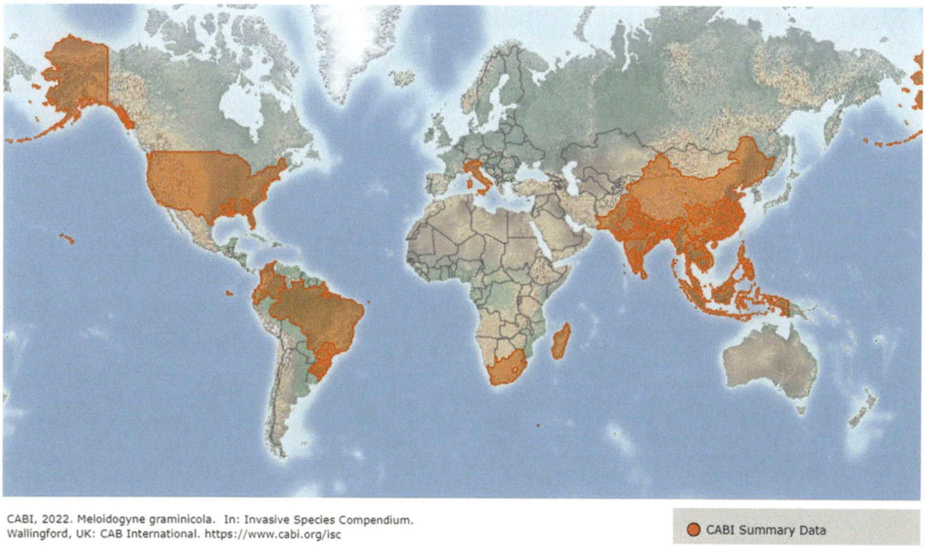

Figure 1.
Global distribution of root-knot nematode in rice.

Figure 2.
Distribution of rice root-knot nematode in India.

Nadu, West Bengal and Uttar Pradesh [29–31]. The nematode was reported on irrigated rice in Andhra Pradesh, Telangana and Karnataka [32]. Severe infestation of *M. graminicola* occurs in upland and sometimes transplanted rice in north-eastern states, West Bengal, Odisha, Bihar, eastern Uttar Pradesh, Chhattisgarh, parts of Madhya Pradesh and Karnataka, Jammu, Punjab, Himachal Pradesh, Haryana, Delhi and Uttar Pradesh (**Figure 2**) [33–36].

3. Symptomptology of root-knot disease of rice

The sever infestation of *Meloidogyne graminicola* resulted in stunted growth, yellowing and patchiness in nursery as well main field. Under severe condition, reduced tillering, leaf size, poor and earhead emergence no earheads may be produced. It results in reduced grain yields. Under below ground parts of the plant formation of terminal hook or typical ring like spindle or bead/nodule shaped galls on the roots [37] (**Figures 3** and **4**). The infected plants exhibit reduced vigour, yellowing and sometimes curling along the

Figure 3.
Nursery and main field showing uneven yellowish patches and galls on root system.

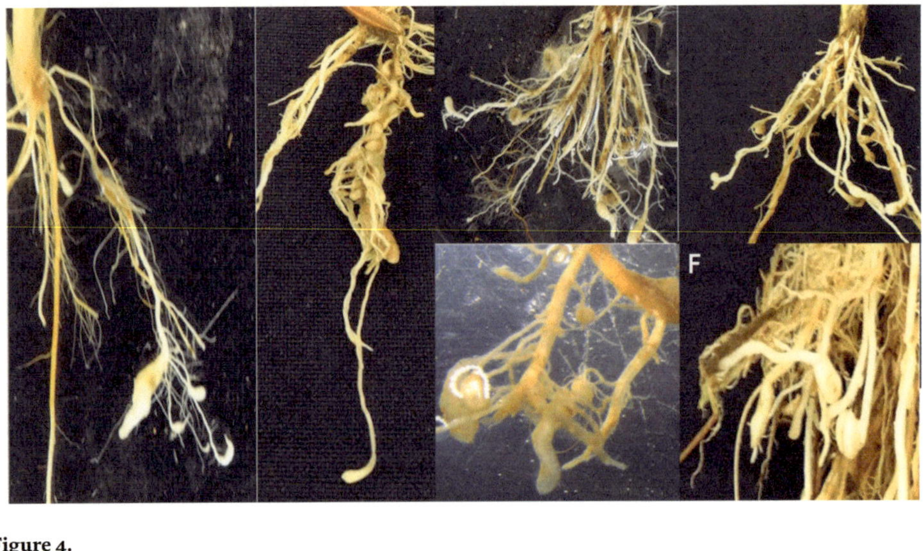

Figure 4.
Symptoms of Meloidogyne graminicola *and different galling pattern.*

midribs. The J_2 (second stage juveniles) cause responsible for induction of feeding sites that results in and hypertrophy and hyperplasia vascular tissues and cortex cells respectively [38]. High initial population of *M. graminicola* causes seedling wilt along with severe reduction in growth parameters, whereas, low population cause only reduction in growth parameters [39]. Characteristic hook like galls on roots, newly emerged leaves appear distorted and crinkled along the margins, stunting, chlorosis, heavily infected plants flower and mature early [24]. Yellowing, dwarfing and gall formation on the roots of rice plants. The degree of symptom manifestation differs with time of infection, age of the plants and load of inoculums [40]. The main symptoms caused by *Meloidogyne graminicola* are yellowing, stunting and gall formation on the roots of rice plants. The degree of symptom manifestation varies with inoculums load, time of infection, age of the plants, *etc*.

4. Biology of *M. graminicola*

M. graminicola is a meiotic parthenogen, with a haploid chromosome number of 18. *M. graminicola* completes its life cycle in 26–51 days in different periods of the year [41]. The J_2 stage requires 16 days at 26°C for converting into matured female and about 8–11 days at 26°C for egg to J_2 (second stage juvenile) [42, 43]. Studied the life cycle of *M. graminicola* and they found 24 days is enough to complete its life cycle. They reported that adult male and females were observed on day 10 and egg laying and release of juveniles were first observed on day 20 and 24, respectively [44], reported that life cycle of *M. graminicola* required 15–20 days to complete its life cycle in rice during different months in eastern Uttar Pradesh condition where temperature usually ranges between 22 and 40° C. Various scientists have been studied the life cycle of *M. graminicola.* According the study conducted by [45] the females of

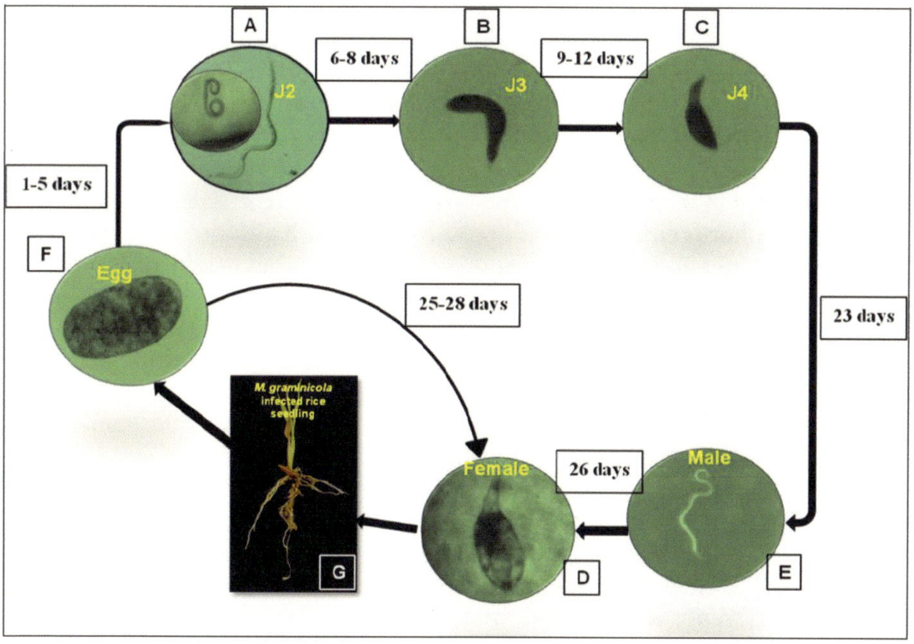

Figure 5.
Life cycle of rice root-knot nematode (M. graminicola).

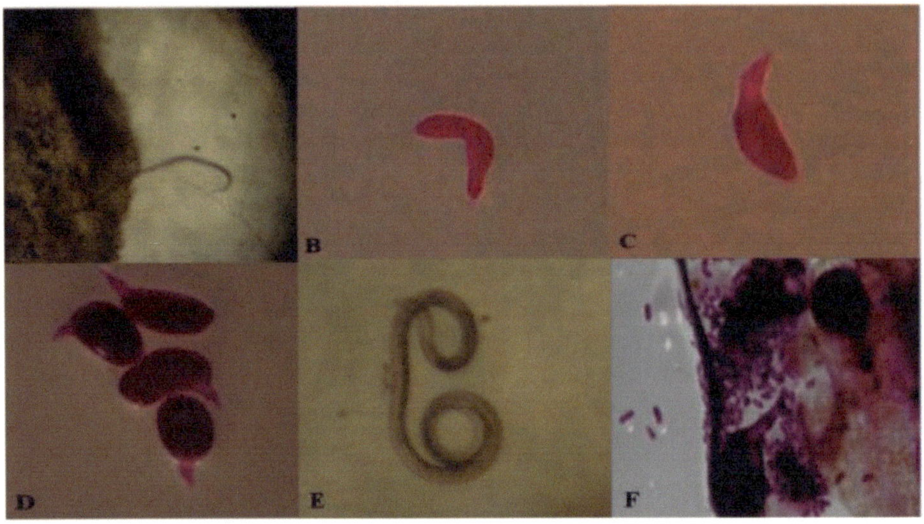

Figure 6.
Biology of M. graminicola, (A) Infective J2, (B) J3, (C) J4, (D) Adult females (E) Adult male and (F) Female and egg.

M. graminicola lay about 250–300 eggs in an egg sac inside the root tissues and the total duration of life cycle was about 25–28 days (**Figures 5** and **6**).

5. Survival strategy

Survival of *M. graminicola* is mainly depends on the edaphic factors of soil and host factor. Various researchers have been reported that *M. graminicola* survival rate was more in moist and wet soil than air dried soil. Similarly the hatching of J_2 is highly inhibited by soil factor i.e., too wet and too dry soil [46, 47]. Application of nitrogen and phosphorus reduces the nematode population as compared to control plots (no fertiliser or compost). In contradict [48], found addition of nitrogen up to 40 kg/ha to the soil resulted in increased reproduction of *M. graminicola*. Application of additional phosphorus either alone or in combination with nitrogen also favoured nematode development. Rao andl Israel [49] reported maximum hatching of eggs of *M. graminicola* in water at 25°C and 30°C. At 15°C and 35°C hatching was reduced and at 20°C it was slightly less than that at 25°C [48], noticed that the juveniles entry in to the host was highest in soils with 32% moisture; similarly egg production were highest at 20–30% soil moisture and greatest juvenile invasion was observed at pH 3.5. Sandy or loamy, laterite soils or recent alluvial soils favour development of the nematode [49]. It has been observed that waterlogged condition in the direct seeded rice or transplanted crop had no detrimental effects on the survival of the endoparasitic stages [50]. Temperature of 22–29°C was found to be suitable for the prevalence of the nematode [32, 41]. Reported higher damage of root-knot in unflooded condition compared to the flooded condition at both ambient (30–40°C) and at high (40–45°C) temperature [51], reported that application of ammonia-based nitrogen fertiliser to the rice nursery bed may interfere with nematode attraction and thus reduce invasion, and the application of chemical nitrification inhibitors to rice nursery beds may decrease nematode invasion [52], studied the role of nutrient on infestation of

M. graminicola and reported that roots supplied with a 100-fold lower supply of calcium nitrate (0.1 mM $Ca(NO3)^2$) showed a higher level of nematode infection with higher root knot index [53], reported that at pH range between 6.5 and < 8.5 in the districts viz., Davanagere, Dakshina Kannada, Udapi, Uttar Kannada, Mysore, Kodagu and Haveri having moderate infection of *M. graminicola* with root-knot index of 3.0 [54], observed the higher nematode population at pH range 4.96–5.96 and soil organic carbon ranging from (1.50 to 1.59) and sandy loam soil in places *viz.*, Chikadadkatte of Davanagere, Tuduru, Beguvalli, Megaramakki and Kadinabellu of Chickmagaluru district (586.33–841.00 J2/100 cc of soil).

6. Management

6.1 Host plant resistance

Resistant cultivars play an important role in sustainable management of nematodes. Since, they are very cheaper and economically friendly [55]. Exploration of resistance sources to *M. graminicola* in rice must be performed under favourable climatic conditions for maximum damage by this nematode [56, 57], screened 414 rice cultivars under artificially inoculating 15-day-old pot-grown seedlings with 100 second stage juveniles. Out of 414 cultivars, only two entries from breeding lines, 127-28-1-1-1 &183-6-1-1-3 were found resistant with score 2 [58], reported rice varieties Loknath 505 and M-36 resistant to the rice root-knot nematode, *M. graminicola* from Allahabad [59], evaluated 50 basmati rice germplasms against *M. graminicola* at crop research centre SVPUAT, Meerut. Out of 50 rice germplasms tested, the germplasms such as Pusa 1637-18-7-6-20 was found to be resistant with scale 2, while, 2 germplasms Shaan (Hybrid) and UPR 3805-12-2-7 were found to be susceptible with scale 4 [60], tested 87 cultivars of rice and 59 cultivars of wheat against *M graminicola*. The study revealed that two rice cultivars Achhoo and Naggardhan and two wheat cultivars HS 295 and VL 829 as resistant with 2 score. Out of 145 local cultivars, 32 and 45 local cultivars were found to be highly resistant and resistant respectively against *M graminicola* [61, 62], evaluated 20 rice genotypes against *M.graminicola.* Out of 20, only one genotype KMP-179 was found to be highly resistant which recorded least root-knot index (1.6) [63]. Screened 136 rice varieties, out of which, Zhonghua 11 (aus), Shenliangyou 1 (hybrid *aus*) and Cliangyou 4418 (hybrid *indica*) were highly resistant to *M. graminicola* under both pot and field conditions.

6.2 Biological control

Biological control is one of the component in Integrated nematode management system it improves the sustainable management of the nematode, soil health and quantity of rice. Bio-control being a ecofriendly and possible alternative to chemicals and safe for disease management *i.e.,* nematode management, perhaps, it is free from toxic residual effects. There are various fungal and bacterial bioagents against *M. graminicola* and their application sequels in significant decrease in the nematode damage [64]. Application of the biocontrol agents such as *Aspergillus niger*, *Pochonia chlamydosporia* and *Pseudomonas fluorescens* proved to be more effective, and significantly reduced the nematode disease in rice [5]. Application of consortium of bio-control agents, *P. fluorescens*@20gm/sq.m + *Trichoderma harzianum* @ 20 g/sq. m was reported to be one of the best treatment in reducing the *M. graminicola* population and increase in rice growth parameters [65]. Reported the application of combination of neem cake+ Vermicompost + Trichoderma spp. was found superior in comparison to other treatment in suppression of root gall formation on rice root in field [66], reported the application of mixtures of *P. fluorescens* strains with PF1 + TDK1 + PY15 signficantly reduced *M. graminicola* infestation when applied through seed treatment.

6.3 Cultural control

Burning of 15 cm deep rice hulls significantly reduce *M. graminicola* populations in the soil [67]. Summer ploughing and puddling of main fields before transplanting and Crop rotation with non-host crops like jute, mustard, chickpea and resistant varieties reduces *M. graminicola* infestation [33]. Crop rotation with non-host plants such as sweet potato, cowpea, sesame, castor, sunflower, soybean, turnip, cauliflower, jute, mustard and chickpea for at least 12 months are recommended to help manage rice root-knot nematode [26].

6.4 Chemical control

Root-dip and soil application of phorate 10G (25 mg a.i. /pot), carbofuran 3G (83.3 mg a.i./pot), carbosulfan 20EC (5 μL/pot) and chlorpyriphos 20 EC (6.25 μL/pot) reduces root-knot infestation in rice [68, 69], reported that application of Phosphonothioate 10G at 1 kg a.i./ha at 7 days prior to uprooting plus main field application at 45 days after transplanting at 1 kg a.i./ha resulted in maximum reduction in population of *M. graminicola* and increase in yield. Soil application and root dip of *P. fluorescens* or *T. harzianum* + carbofuran was found most effective and suppressed the gall formation (40–46%) and increased the yield up to 37–42% [70, 71], reported the fumigation of 1, 3-dichloropropene can help to reduce the number of nematodes before planting [72], adopted the Integrated Nematode Management Technology for the management of *M. graminicola* the results indicated that reduction in nematode population from 320 J_2/200 cc soil to 135 J_2/200 cc soil was recorded in cabrofuran (0.3 g) a.i/m^2 with more yield (4.72 tonnes/ha), followed by bioagents 165 J_2/200 cc soil (*Pseudomonas fluorescence* at 20 g/m^2) with 4.67 tonnes/ha yield and 192 J_2/200 cc soil (*Trichoderma viride* at 4 g/200 cc of soil) with 4.29 tonnes/ha respectively [73], investigated application of bioagents along with nematicides found to be better in managing the infestation and increasing the yield of rice. Soil application of *P. fluorescens* at 20 g/m^2 + cabrofuran (0.3 g a.i/m^2) found to be good in increasing growth of plant parameters *viz.*, plant height (83.26 cm), root length (20.60 cm), maximum grain yield (44.1 q/ha) and least nematode population (132.67/200 g soil) with reduction of 79.34% nematode population.

7. Conclusion

Of late, *Meloidogyne graminicola* has becoming a major menace in rice. Due to change in global temperature, wider distribution, adoptability nature under adverse condition, wider host range and different survival stratagies denotes it has becoming a major menace in rice. Since there are many management practices available for nematode management but, sustainable management can be achieved by integration of cultural, physical, biological, host-plant resistance and usage of chemicals.

The attention must be given to identify the different species involved in root-knot association, development of resistant varieties and development new nematicide for the management of this notorious pest in rice.

Author details

Hosahalli Buthanna Narasimhamurthy[1*], Mukesh Sehgal[2] and R. Ganesha Naik[1]

1 Department of Plant Pathology, Keladi Shivappa Nayaka University of Agricultural and Horticultural Sciences, Shivamogga, Karnataka, India

2 ICAR-National Centre for Integrated Pest Management, New Delhi, India

*Address all correspondence to: namunema24@gmail.com

References

[1] Ngangkham U, Parida SK, De S, Kumar KAR, Singh AK, Singh NK, et al. Genetic markers for wild abortive (WA) cytoplasm based male sterility and its fertility restoration in rice. Molecular Breeding. 2010;**26**:275-292

[2] Marathi B, Jena KK. Floral traits to enhance outcrossing for higher hybrid seed production in rice: Present status and future prospects. Euphytica. 2015;**201**:1-14

[3] Ravindra H, Sehgal M, Dinesh KM, Manu TG, Murali R, Narasimhamurthy HB, et al. Survey and digital mapping of rice root-knot nematode incidence in Karnataka, India. Indian Journal of Nematology. 2014;**44**(1):27-31

[4] Narasimhamurthy HB, Ganesha Naik R, Sehgal M, Malik M. Integrated management of rice diseases. Journal of Advanced Research in Agriculture Science & Technology. 2021;**4**(2):13-24

[5] Sehgal M, Narasimhamurthy HB. Field management of rice root-knot nematode, *Meloidogyne graminicola* using bio-agents: An environmental benign approach. Indian Journal of Nematology. 2021;**51**(2):181-191

[6] Ravindra H, Mukhesh S, Narasimhamurthy HB, Jayalakshmi K, Imran Khan HS. Rice root-knot nematode (*Meloidogyne graminicola*) an emerging problem. International Journal of Current Microbiology and Applied Sciences. 2017;**6**(8):3143-3171

[7] Jones JT, Haegeman A, Danchin EG, Gaur HS, Helder J, Jones MG, et al. Top 10 plant parasitic nematodes in molecular plant pathology. Molecular Plant Pathology. 2013;**14**:946-961

[8] Siddiqi MR. Tylenchida-parasites of plants and insects. In: Published by Commonwealth Institute of Parasitology. Willingford, UK: Commonwealth Agricultural Bureaux; 1986. p. 586

[9] Jain RK, Mathur KN, Singh RV. Estimation of losses due to plant parasitic nematodes on different crops in India. Indian Journal of Nematology. 2007;**37**:219-220

[10] Narasimhamurthy HB, Ravindra H, Sehgal M. Management of rice root-knot nematode, *Meloidogyne graminicola*. International Journal of Pure and Applied Biosciences. 2017;**5**(1):268-276

[11] Ravindra H, Sehgal M, Narasimhamurthy HB, Jayalakshmi K, Soumya DM. Status of *Meloidogyne graminicola* under different organic carbon regimes of rice growing districts of Karnataka. Current Nematology. 2015;**26**(1,2):7-12

[12] Prot JC, Villanueva LM, Gergon EB. The potential of increased nitrogen supply to mitigate growth and yield reductions of upland rice cultivar UPL Ri-5 caused by *Meloidogyne graminicola*. Fundamentals Applied Nematolology. 1994;**17**(5):445-454

[13] De Waele D, Elsen A. Challenges in tropical plant nematology. Annual Review of Phytopathology. 2007;**45**:457-485

[14] Bridge J, Plowright RA, Peng D. Nematode parasites of rice. In: Luc M, Sikora RA, Bridge J, editors. Plant Parasitic Nematodes in Subtropical and Tropical Agriculture. Wallingford, Oxfordshire: CABI Bioscience; 2005. pp. 87-130

[15] Soriano IRS, Prot JC, Matias DM. Expression of tolerance for

Meloidogynegraminicola in rice cultivars as affected by soil type and flooding. Journal of Nematology. 2000;**32**:309-317

[16] Song ZQ, Zhang DY, Liu Y, Cheng FX. First report of *Meloidogyne graminicola* on rice (*Oryza sativa*) in Hunan Province, China. Plant Disease. 2017;**12**:2153

[17] Tian ZL, Maria M, Barsalote EM, Castillo P, Zheng JW. Morphological and molecular characterization of the rice root-knot nematode, *Meloidogyne graminicola*, Golden and Birchfield, 1965 occurring in Zhejiang, China. Journal of Integrative Agriculture. 2018;**17**:2724-2733

[18] Prasad JS, Panwar MS, Rao YS. Screening of some rice cultivars against the root-knot nematode *Meloidogyne graminicola*. Indian Journal of Nematology. 1986;**16**:112-113

[19] Ravindra H, Narasimhamurthy HB, Sehgal M, Girijesh GK. Investigation of rice root-knot nematode and weeds interactions in rice agro ecosystem. Journal of Crop and Weed. 2017;**13**(1):157-163

[20] Golden AM, Birchfield W. *Meloidogyne graminicola* (Heteroderidae), a new species of root-knot nematode from grass. Proceedings of the Helminthological Society of Washington. 1965;**32**:228-231

[21] Bridge J. Nematode parasites of rice. In: Luc M, Sikora RA, Bridge J, editors. Plant-parasitic Nematodes in Subtropical and Tropical Agriculture. UK: CAB International; 1990. pp. 75-107

[22] Munir A, Bridge J. Rice root-knot nematode *Meloidogyne graminicola* Golden and Birchfield, 1965 from rice in Pakistan. Pakistan Journal of Nematology. 2003;**21**(2):133-135

[23] Bridge J, Luc M, Plowright RA, Peng D. Nematode parasites of rice. In: Luc M, Sicora RA, Bridge J, editors. Plant Parasitic Nematodes in Tropical and Subtropical Agriculture. UK: CABI; 2005. pp. 87-130

[24] Dutta TK, Ajoy K, Ganguly AND, Gaur HS. Global status of rice root-knot nematode, *Meloidogyne graminicola*. African Journal of Microbiology Research. 2012;**6**:6016-6021

[25] Kyndt T, Fernandez D, Gheysen G. Plant-parasitic nematode infections in rice: Molecular and cellular insights. Annual Review of Phytopathology. 2014;**52**:135-153

[26] Mantelin S, Bellafiore S, Kyndt T. *Meloidogyne graminicola*: A major threat to rice agriculture. Molecular Plant Pathology. 2017;**18**:3-15

[27] Sacchi S, Torrini G, Marianelli L, Mazza G, Fumagalli A, Cavagna B, et al. Control of *Meloidogyne graminicola* a root-knot nematode using rice plants as trap crops: Preliminary results. Agriculture. 2021;**11**:37

[28] Patnaik NC. Pathogenicity of *Meloidogyne graminicola (Golden and Birchfield 1965) in rice (Abstr.)*. New Delhi: All India Nematology Symposium; 1969, 1969. p. 12

[29] Prasad JS, Panwar MS, Rao YS. Nematode problems of rice in India. The Pest Management. 1987;**33**:127-136

[30] Mukhopadhyay AK, Khan MR. Distribution of *Meloidogyne graminicola* and other nematodes in rice-based cropping system in West Bengal. In: Nematode Pests in Rice-wheat-Legume Cropping Systems: Proceedings of Review and Planning Meeting and Training Workshop. New Delhi: Division of Nematology, Indian Agricultural Research Institute; 1999. pp. 20-21

[31] Somasekhar N, Prasad JS. Root-knot nematode *Meloidogyne graminicola*-An emerging threat to rice cultivation. DRR Newsletter. 2009;7:3-4

[32] Prasad JS, Vishakanta, Gubbaiah. Outbreak of root-knot nematode, *Meloidogyne graminicola* disease in Mandya district, Karnataka State and farmers' perceptions. In: National Congress on Centenary of Nematology in India-Appraisal and Future. New Delhi: IARI; 2001

[33] Pankaj Sharma HK, Prasad JS. The rice root-knot nematode, *Meloidogyne graminicola:* an emerging problem in rice-wheat cropping system. Indian Journal of Nematology. 2010;**40**:1-11

[34] Gaur HS, Singh J, Sharma SN, Chandel ST. Distribution and community analysis of plant-parasitic nematodes in rice-growing areas of Haryana, India. Annals of Plant Protection Sciences. 1996;**4**:115-121

[35] Pankaj Sharma HK, Gaur HS, Singh AK. Effect of zero tillage on the nematode fauna in a rice-wheat cropping system. Nematologia Mediterranea. 2006;**34**:175-178

[36] Singh SK, Singh KP. A first record of root-knot nematode (*Meloidogyne graminicola*) on wheat in Uttar Pradesh, India. Journal of Mycology and Plant Pathology. 2009;**39**:340-342

[37] Narasimhamurthy HB, Ravindra H. Invasion studies and symptomatology of rice root-knot nematode, *Meloidogyne graminicola*. Advances in Life Sciences. 2016;**5**(22):10408-10411

[38] Jene RN, Rao YS. Root-knot nematode resistance in rice. In: Breeding Researches in Asia and Oceana. New Delhi: Proc. Second General Congr. SABRAO; 1973. pp. 108-109

[39] Plowright R, Bridge J. Effect of *Meloidogyne graminicola* (Nematoda) on the establishment, growth and yield of rice cv. IR36. Nematology. 1990;**36**:81-89

[40] Vinod U, Nitish RB, Neelam, Sajeesh PK. *Meloidogyne graminicola* (Golden and Birchfield): Threat to Rice Production. Research Journal of Agriculture and Forestry Sciences. 2014;**2**(3):31-36

[41] Rao YS, Israel P. Life history and binomics of *Meloidigyne graminicola*, the rice root-knot nematode. Indian Phytopathology. 1973;**26**(2):333-340

[42] Yik CP, Birchfield W. Additional knowledge on the biology and hosts of *Meloidogyne graminicola*. American Phytopathological Society. 1977;**4**:231

[43] Dabur KR, Taya AS, Bajaj HK. Life cycle of *Meloidogyne graminicola* on paddy and its host range studies. Indian Journal of Nematology. 2004;**34**:80-84

[44] Singh KP, Jaiswal RK, Kumar N, Kumar D. Biomass of nematode and associated roots: A determinant of symptom production in root knot disease of rice (*Oryza sativa* L.). Journal of Phytopathology. 2006;**154**(11-12):676-682

[45] Narasimhamurthy HB, Ravindra H, Sehgal M, Rani N, Ekabote SD, Ganapathi. Biology and life cycle of rice root-knot nematode (*Meloidogyne graminicola*). Journal of Entomology and Zoology Studies. 2018;**6**(1):477-479

[46] Gaur HS. Root-knot Disease of Rice and Wheat: Problem and Management. New Delhi: lARJ; 2003. p. 20

[47] Pokharel RR. Reaction of commonly grown rice and wheat varieties in Nepal to *Meloidogyne graminicola*. Journal of Nematology. 2004;**36**:342

[48] RAO YS, Israel P. Studies on nematodes of rice and rice soils. V. Influence of soil chemical properties on the activity of *Meloidogyne graminicola,* the rice root-knot nematode. Oryza. 1971;**8**(2):33-38

[49] Rao YS, Israel P. Studies on nematodes of rice and rice soils. VI. Influence of soil temperature on the prevalence and incidence of *Meloidogyne graminicola,* the rice root-knot nematode. Oryza. 1971;**8**(2):47-51

[50] Prasad JS, Panwar MS, Rao YS. Occurrence of root knot-nematode, *Meloidogyne graminicola* in semideepwater rice. Current Science. 1985;**54**:387-388

[51] Patil J, Miller AJ, Gaur HS. Effect of nitrogen supply form on the invasion of rice roots by the root-knot nematode, *Meloidogyne graminicola*. Nematology. 2013;**15**(4):483-492

[52] Jagadeesh P, Stephen JP, Keith GD, Gaur HS, Miller AJ. Effect of root nitrogen supply forms on attraction and repulsion of second-stage juveniles of *Meloidogyne graminicola*. Nematology. 2013;**15**:469-482

[53] Ravindra H, Sehgal M, Narasimhamurthy HB, Soumya DM, Imran Khan HS. Survival strategies of *Meloidogyne graminicola* under different soil pH regimes of rice growing districts of Karnataka. Current Nematology. 2014;**25**(1,2):25-29

[54] Narasimhamurthy HB, Ravindra H, Sehgal M. Survey on incidence of rice root-knot nematode (*Meloidogyne graminicola*) in Shivamogga and Davanagere Districts of Karnataka, India. Advances in Life Sciences. 2016;**5**(14):5525-5529

[55] Amoussou PL, Ashurt J, Green J, Jones M, Koyama M, Snape JTW, et al. Broadly based Resistance to Nematodes in the Rice and Potato Crops of Subsistence Farmers. DFID Plant Sciences Research Programme Annual Report; 2004. pp. 9-14

[56] Tandingan IC, Prot JC, Rowulo GD. Influence of water management on tolerance of rice cultivars for *Meloidogyne graminicola*. Fundamental of Applied Nematology. 1996;**19**:189-192

[57] Berliner J, Somnath SP, Mishra C, Jena M, Singh ON. Screening of rice germplasm lines against rice root-knot nematode *Meloidogyne graminicola*. ICAR-CRRI, Cuttack, Oryza. 2014;**51**(2):177-178

[58] Hassan MG, Pant VR, Devi LS. Infestation of rice root-knot nematode (*Meloidogyne graminicola*) associated with different varieties of rice in Allahabad District of Uttar Pradesh, India. Indian Journal of Nematol. 2004;**34**:227

[59] Rana R, Khilari K, Singh G, Mukesh SJ, Bansal A, Dwivedi A. Screening of basmati rice germplasms against rice root-knot nematode, *Meloidogyne graminicola*. Journal of Pure and Applied Microbiology. 2016;**10**(3):2261-2265

[60] Srivastava A, Rana V, Rana S, Singh D, Singh V. Screening of rice and wheat cultivars for resistance against root-knot nematode, *Meloidogyne graminicola* (Golden and Birchfield) in rice-wheat cropping system. Journal of Rice Research. 2011;**4**:1-2

[61] Ravindra H, Sehgal M, Narasimhamurthy HB, Khan I, Shruthi. Evaluation of rice landraces against rice root-knot nematode, *Meloidogyne graminicola*. African Journal of Microbiology Research. 2015;**9**(16):1128-1131

[62] Narasimhamurthy. Investigations on root-knot nematode (*Meloidogyne*

graminicola Golden and Birchfield, 1965) of rice (*Oryza sativa* L.). *M.Sc. (Agri.) Thesis*. Shivamogga: University of Agricultural and Horticultural Sciences; 2014. p. 81

[63] Zhan L-p, Ding Z, Peng D-l, Peng H, Kong L-a, Liu S-m, et al. Evaluation of Chinese rice varieties resistant to the root-knot nematode *Meloidogyne graminicola*. Journal of Integrative Agriculture. 2018;**17**(3):621-630

[64] Khan MR, Haque Z, Ahamad F, Zaidi B. Bio-management of rice root-knot nematode *Meloidogyne graminicola* using five indigenous microbial isolates under pot and field trials. Journal of Applied Microbiology. 2021;**130**(2):424-438

[65] Kumar D, Khilari K, Kumar N, Jain SK. Integrated disease management of rice root knot nematode (*Meloidogyne graminicola*) through organic amendments, Trichoderma spp. and Carbofuran. Journal of Pharmacognosy and Phytochemistry. 2017;**6**(6):2509-2515

[66] Seenivasan N, David PMM, Vivekanandan P, Samiyappan R. Biological control of rice root-knot nematode, *Meloidogyne graminicola* through mixture of *Pseudomonas fluorescens* strains. Biocontrol Science and Technology. 2012;**6**:611-632

[67] Gergon EB, Miller SA, Davide RG, Opina OS, Obien SR. Evaluation of cultural practices (surface burning, deep ploughing, organic amendments) for management of rice root-knot nematode in rice-onion cropping system and their effect on onion (*Allium cepa* L.) yield. International Journal Pest Management. 2001;**47**(1):265-272

[68] Khan MR, Bushra Z, Ziaul H. Nematicides control rice root-knot, caused by *Meloidogyne graminicola*. Phytopathologia Mediterranea. 2012;**51**:298-306

[69] Das D, Choudhury BN. Chemicals for the management of rice root-knot (*Meloidogyne graminicola*) and rice root (*Hirschmanniella oryzae*) nematodes. Indian Journal of Nematology. 2012;**42**(2):107-110

[70] Haque Z. Development of integrated nematode management module for rice root-knot disease caused by *Meloidogyne graminicola:* A success story for endemic area. In: Conference Proceeding: National Symposium on Nematode: A Friend and Foe to Agri-horticultural Crops. (H.P.), India: at solan; 2013

[71] Dutta TK, Ajoy K, Ganguly, Gaur HS. Lobal status of rice root-knot nematode, *Meloidogyne graminicola*. African Journal of Microbiology Research. 2012;**6**:6016-6021

[72] Somasekhara Y, Sehgal M, Ravichandra NG, Siddegowda DK, Jain RK, Mahadevu P, et al. Validation and economic analysis of adaptable integrated management technology against root-knot nematode (*Meloidogyne graminicola*) in rice (*Oryza sativa*) with farmers' participatory approach. Indian Journal of Agricultural Sciences. 2012;**82**:442-444

[73] Narasimhamurthy HB, Ravindra H, Divya Bharathi AR, Imran Khan HS, Saritha AG. Survey and integrated management of rice root-knot nematode, *Meloidogyne graminicola*. In: National Symposium on Recent Trends in Plant Pathological Research and Education. UAS, Raichur: Indian Phytopathological Society (S-Z); 2016. p. 91

Chapter 2

Breeding Strategies for Improvement of Drought Tolerance in Rice: Recent Approaches, and Future Outlooks

Banoth Madhu, Bhimireddy Sukrutha, Nunavath Umil Singh and Govada Venkateswarao

Abstract

Rice production is severely limited by drought stress, which causes significant monetary losses. The global climate change is turning into a more significant problem. Enhancing agricultural yield in the drought-prone rainfed areas has become critical in light of the current and projected global food demand. There is a need for rice varieties with drought tolerance in order to achieve the production objective from rainfed areas, and genetic improvement for drought tolerant should be a high priority issue of study in the future. The intricate structure of breeding for drought-tolerant rice varieties makes it a challenging endeavour, and multigenic regulation of drought-tolerant features would be a significant roadblock for the ongoing study. In this chapter, we discussed on the recent crop improvement program for the development of drought-tolerant rice varieties and highlighted the most recent advancements through conventional to molecular breeding level for adaption of cultivars against drought tolerance in rice under different agro-climatic conditions.

Keywords: adaptation, climate change, drought stress, improvement, rice

1. Introduction

Globally, more than one-third of the world's population consumes rice e (*Oryza sativa* L.) as a main staple meal, and a large majority of people, particularly in Asian countries, rely on it for ~80% of their daily caloric needs [1, 2]. In terms of rice production and consumption, Asia is in first place (FAO report, 2020–2021). The tiny root system, thin cuticular wax, and quick stomata closure of rice, however, make it one of the plants most vulnerable to drought [3–5]. A prerequisite for the effort to achieve self-sufficiency in rice production by 2050 is the creation of high yielding rice varieties with a high level of tolerance and resistance to both biotic and abiotic stressors under adverse climatic conditions [6, 7]. Stress may result from biotic causes like the prevalence of pests, insects, and diseases or abiotic factors such heavy metal toxicity, flooding, salinity, drought, high temperatures, and air pollution, among others. Drought is one of the most destructive abiotic elements [8, 9]. Depending on the plant's stage of growth, drought stress can result in complete yield losses. Yield losses must be kept to a minimum to aid impoverished rice farmers in emerging nations and ensure food sustainability for the world's expanding population [9].

The two most significant limiting factors for the low production of rice worldwide are the escalating severity of droughts and the scarcity of high-yielding genotypes that can be grown in drought-prone environments [8]. Due to a lack of suitable rice cultivars and farming methods, rice cultivation is seasonal. Rice cultivation is impacted by the decrease in water supplies brought on by the depletion of important groundwater resources. Due to their immobility, plants have very little chance of escaping the drought state [10]. Severe drought stress can be damaging to plant development at all stages. Low reproductive success for many plant species is caused by the consequences of water deficiency during the reproductive development stage, which can result in male sterility and embryo abortion shortly after pollination [4, 11]. Therefore, understanding how plants respond to the stress becomes vital and primary to designing plants that are resistant to such stress.

Genotype, environment, and the interaction between genotype and environment all have a role in how a plant grows and develops. The biochemical activities that are influenced by environmental influences are also necessary for development [12]. Plants become stressed when environmental conditions are not optimal, which negatively impacts their production, growth, and development. There are two distinct categories of drought conditions: terminal [13] and intermittent [9]. A terminal drought state is brought on by a reduction in the amount of water that is accessible to plants, which causes extreme drought stress and the eventual death of the plant. However, intermittent drought conditions, which happen once or repeatedly during planting seasons, cause plant growth to suffer during the periods of insufficient irrigation. Intermittent drought conditions, in contrast to terminal drought stress, are typically not fatal. Plant survival and ability to retain function during intermittent and terminal drought conditions are key components of drought tolerance or resistance mechanisms [4, 14, 15].

Over the past few decades, study on drought has been increasingly important due to both its rising frequency and its significance for crop output. Nevertheless, it has been difficult to examine drought responses due to the quantitative and complicated character of the drought-tolerant trait [16]. Rice productivity can be increased in a sustainable and economically feasible way by breeding rice cultivars that are tolerant to drought stress [9, 15]. Researchers have tried to breed for drought-tolerant rice plants in the past, but because there aren't many donors with a high level of drought tolerance, progress is being made slowly. Only a few drought-tolerant variants have yet been identified after screening thousands of samples of germplasm for drought resistance in different parts of the world [17]. The main causes of the limited success are the absence of really drought-tolerant genotypes and the lack of appropriate screening techniques [7, 15]. Nearly 1000 Gene bank accessions originating from 47 different countries were examined for drought tolerance over the course of the previous two decades by researchers at the International Rice Research Institute (IRRI), in the Philippines [2, 18] they have discovered 65 more *aus* or *indica* accessions that can withstand drought [19]. In terms of *aus* accessions, the majority of drought-tolerant varieties are from Bangladesh (19), followed by India (7), while the most of drought-tolerant varieties are from India (16), Bangladesh (3), and Sri Lanka (3) [2]. The use of these rice accessions in next crop improvement projects requires molecular genetics and characterisation for drought tolerance. The most promising sources of genes related to drought that can be employed in the creation of contemporary crop varieties are those cultivars that display great drought resistance [5].

Therefore, one of the most crucial phases in the development of the drought-tolerant crop is knows how plants react to drought stress. The goals of this chapter were (i) to explain how drought stress affects rice plants and to highlight current developments in rice's physiological, biochemical, and molecular adaptation to drought tolerance (ii) to describe the current process for creating a long-lasting rice variety that is drought-resistant through conventional breeding and the application of biotechnological tools, and (iii) to conduct a thorough analysis of the information that is currently available on drought-resistant genes/QTLs, QTL analysis, gene introgression, and marker-assisted selection.

2. Mechanisms of drought stress and their responses to drought stress in rice

The word "stress" is frequently interpreted physically, as a reaction to various circumstances. Stress is typically an alteration of physiological conditions brought on by elements that seek to compromise the plant's stability [20]. Low or no precipitation is a climatic characteristic of drought. The majority of the time, drought pressures develops when there is little water in the soil and a constant loss of water through evaporation and transpiration [6]. The term "drought tolerance" refers to a plant's ability to produce its highest economic yield when water is scarce [21]. It is a complex trait depends on the action and interaction of different morphological, biochemical, and physiological responses are some of the mechanisms that are influenced by genetic variables at various stages are shown in **Figure 1** [22]. According to Kumar et al. [22], "drought escape" is defined as the ability of a plant to complete its life cycle before the development of serious soil water deficits. "Drought avoidance" is the ability of plants to maintain relatively high tissue water potential despite a shortage of soil moisture is shown in **Figure 1** [23].

2.1 Responses of plant morphological traits to drought stress in rice

When rice is subjected to water stress, morphological changes in the early stages of the grain are observed. Normal productivity depends on the timely and ideal establishment of a crop stand. Blighted germination and lowered growth are the primary effects of drought stress [23–25]. Due to the lack of water, severe reductions in seed germination and growth are seen during drought stress [26]. The germination process is significantly impacted by drought because it prevents water intake and weakens seedlings [26]. Drought stress interferes with water balance, impairs membrane transport, disrupts metabolic processes at the cellular level, and reduces ATP synthesis and respiration, which results in poor seed germination [24]. Water stress causes declines in plant height, leaf area, and biomass, according to a number of reports [25, 27, 28]. Due of the low water potential caused by the drought, leaf growth is inhibited [29]. Crops respond by having poor cell development and reduced leaf area due to disrupted water passage from the xylem towards another cell, as well as lower turgor pressure as a result of water shortage [28]. Under drought-stressed conditions, the anatomy of the leaf and its ultrastructure are altered [30]. These modifications include reduced leaf size, fewer stomata, thick cell walls, cutinisation of the leaf surface, and inadequate conducting system development [21].

Other significant characteristics of plants under drought stress include rolling of the leaves and the beginning of early senescence [31]. According to [25, 28] larger flag

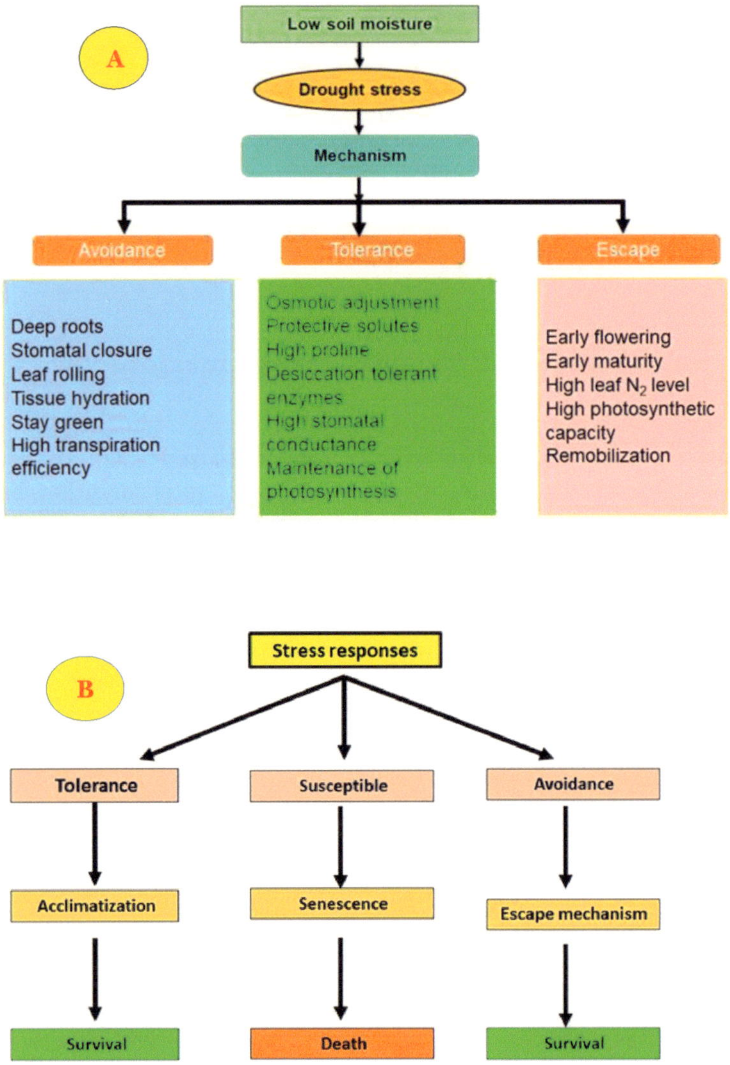

Figure 1.
Mechanisms of drought stress and their responses to drought stress in rice. (A) Different responses and mechanism of the rice plants under drought stress; (B) Plant response mechanisms to drought stress.

leaf area, leaf area index, leaf relative water content, and leaf pigment content have all been used to screen for drought-tolerant varieties of plants. For increasing output while under drought stress, a plant's root properties are essential. The structure and development of rice root system determine crop function under water stress. By using root mass (dry) and length, it is possible to predict rice output under water stress [32]. On the properties of root growth under water stress, a variety of reactions are seen. A rise in the content of abscisic acid in the roots caused to notice that the length of rice roots increased when under drought stress [33]. Generally speaking, rice cultivars having a deep and voluminous root system are more drought-resistant [34, 35]. For rice, genotypes with a deep root system, coarse roots, the ability to produce numerous branches, and a high root-to-shoot ratio are crucial for drought resistance [35]. Under drought stress, the morpho-physiological traits of rice roots significantly influence shoot growth and total grain output [35]. On the other side, morphological adaptations include enlarged roots with longer root lengths, waxy or thick leaf coatings, fewer epithelial cells, delayed leaf senescence, and more green leaf area [3–5].

2.2 Responses of plant physiological traits to drought stress in rice

Water stress from a drought or water shortage affects photosynthesis, one of the key metabolic processes that govern crop growth and yield [29]. When there is not enough water available, the stomata shut, lowering the amount of carbon dioxide that reaches the leaves and causing more electrons to be driven into the reactive oxygen species reaction [34, 36]. The decline of photosynthesis is caused by a number of mechanisms, including stomatal closure, turgor pressure loss, reduced leaf gas exchange, and decreased CO2 uptake, which eventually harms the photosynthetic apparatus [8, 29, 36]. Different representations, such as the water potential of the leaf and relative water content (RWC), can be used to show how a plant and water interact [36]. When plants are under water stress, water consumption efficiency is thought to be a crucial factor in determining their ability to produce. It can be viewed as a strategy for enhancing crop production during drought [34]. RWC is a crucial characteristic of water relations in plants and is regarded as the finest integrated measurement of plant water status because it captures changes in the water potential and turgor potential of the plants [8].

In general, the effects of drought stress include a drop in the water content of plants, a reduction in cell length and growth, the closing of stomata, a decrease in gaseous exchange, and the disruption of enzyme-catalysed activities [2]. Additionally, in times of extreme dryness, photosynthesis and metabolism are severely disrupted, which ultimately results in plant death [37]. When compared to cell divisions, drought stress inhibits cell growth [31]. The several biochemical and physiological processes that are impacted by this restriction on plant growth include ion absorption, respiration, photosynthesis, growth promoters, carbohydrate, source-sink relationships, and nutrient metabolism [38]. Chlorophyll content is increased, osmotic potential is decreased, and harvest index is decreased in cells that have been adapted to withstand dryness. Higher stomatal density and conductance, lower transpiration rates, reduced and early asynchrony between female and male flowering and maturation, and improved production, accumulation, assimilation, and yield partitioning are all characteristics of physiological acclimation [6, 17].

2.3 Responses of plant biochemical traits to drought stress in rice

In response to drought stress, plants accumulate organic and inorganic solutes that lower the osmotic potential in an effort to maintain cell turgor. Osmotic adaptations are provided for the plants through the accumulation of osmoprotectants such as proline, glycinebetaine, and soluble sugar [22, 30]. Drought resistance is improved by protein content and profile, as well as a rise in antioxidant activity for scavenging reactive oxygen species [15]. Improved drought response without lowering yield is achieved via tissue- and time-specific expression of drought-responsive features including abscisic acid, brassinosteroids, and ethylene phytohormone pathways etc., [8].

2.3.1 Osmolyte buildup in a drought-stressed environment in rice

The primary process in plants is osmoregulation, and when turgor declines, osmoprotectants accumulate. Under conditions of water scarcity, accumulation of different osmolytes, such as proline, soluble sugar, phenolic, and total free amino acids, increases and plays a significant role in the ability of plants to withstand drought [16, 31]. Plant cells must detect an above- or below-ground incidence of an imbalance between water loss and water availability before their perception may be translated into a cellular stress signal, which is then used to activate drought resistance systems. Plants, which are sessile organisms, have developed a sophisticated signalling system that uses a variety of primary and secondary signal transduction pathways to spread stress messages throughout the entire plant. Since changes in gene expression frequently involve a mix of hormone signals along with the buildup of additional metabolic products including reactive oxygen species, proteins, and other osmolytes, these pathways contain a variety of signalling molecules [16]. Turgor pressure is maintained in dry conditions by the buildup of organic and inorganic solutes, which reduces the osmotic potential in the cytosol. A kind of osmotic adaptation, this metabolic process is highly dependent on the degree of water stress [16, 31]. Proline [39], sucrose [40], glycine betaine [41], and other solutes etc., build up in the cytoplasm as osmotic adaptation happens, encouraging water uptake.

In plants, proline works as an osmolyte in a variety of harmful situations [42]. There are discrepancies between proline accumulation under stress and normal conditions in rice [43]. Comparing dry conditions to well-watered conditions, the proline buildup rises in all rice cultivars [34]. Higher proline buildup is typically connected with greater resistance to drought, and it aids in maintaining leaf turgor and advancing stomatal conductance [22]. Proline content can therefore be used to screen plants for dehydration using biochemical markers [15, 34]. Carbohydrates/soluble sugars are the structural component that provides the energy needed to support plant biomass. Disaccharides, oligosaccharides, and fructans are primarily three forms of water-soluble carbohydrates that play a critical role in stress tolerance under abiotic stress [40]. The balance of numerous physiological activities, including photosynthesis and mitochondrial respiration, depends heavily on soluble sugars [44]. Since plants use a variety of sugar-based coping mechanisms to adapt to environmental stress, sugars play a variety of roles in plants [41]. The availability of mannitol, sorbitol, and trehalose is crucial for the plant's healthy growth and metabolic operation. Because of the accumulation of soluble sugars that drought causes, the plants are somewhat protected from adverse conditions and even act as osmoprotectants [22, 30, 41].

2.3.2 Antioxidants' function in drought stress

The plants have an antioxidant defence system that protects them from oxidative harm. Both enzymatic and non-enzymatic antioxidants are present. Antioxidants are essential components of plants that scavenge reactive oxygen species (ROS), and rice that expresses them is more drought-tolerant [2, 5, 44]. The most frequent occurrence when there is a drought stress is an imbalance between the generation and quenching of ROS. In rice, a drought-related imbalance in ROS production and quenching can lead to oxidative damage and negatively impact the life cycle of the plant by reacting with proteins, lipids, and deoxyribonucleic acid [44]. Electron leakage to 1 O2 and the subsequent Mehler reaction that produces ROS have a negative impact on photosynthesis. Due to the negative consequences of the photorespiratory pathway during drought, excessive levels of superoxide radical (O_2), hydrogen peroxide (H_2O_2), and hydroxyl radical (OH) are formed [44]. These are

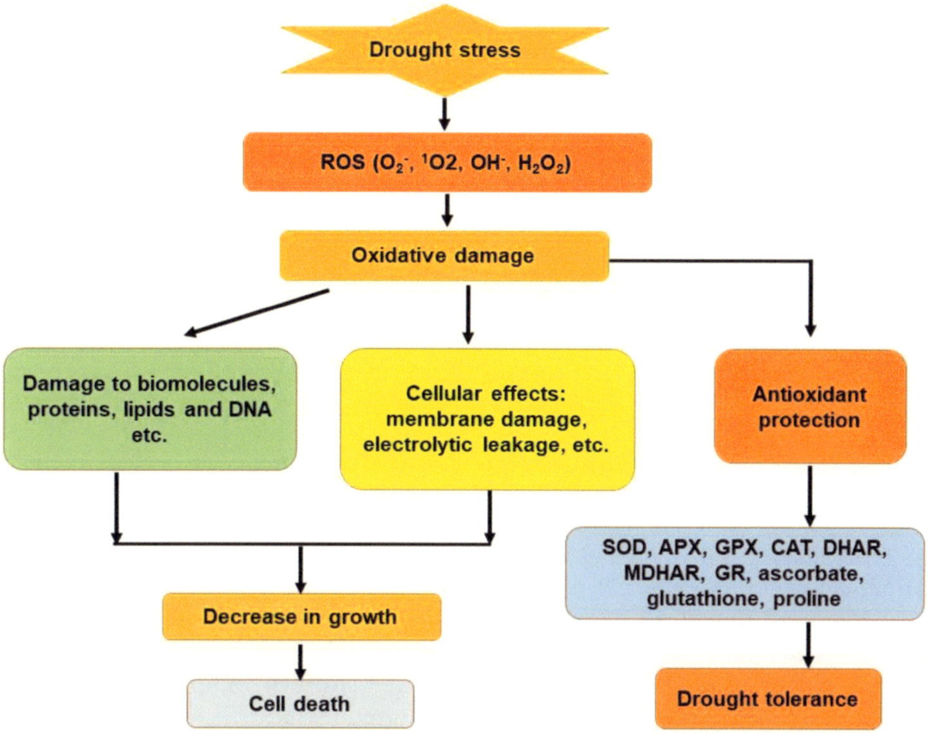

Figure 2.
Schematic representation of reactive oxygen species (ROS) damage and antioxidant protection of rice plants under drought stress [44]. APX, ascorbate peroxidase; CAT, catalase; DHAR, Dehydroascorbate reductase; GR, glutathione reductase; GPX, guaiacol peroxidase; MDHAR, monodehydroascorbate reductase; SOD, superoxide dismutase.

extremely harmful radicals that cause cellular death by causing lipid peroxidation, protein, and membrane damage to a variety of cell components under drought stress [44]. Therefore, reducing excessive ROS production or increasing antioxidant activity in rice organs is the most effective strategy to improve rice's ability to withstand drought. **Figure 2** illustrates the mode of ROS formation, harmful effects of oxidative stress, cell damage that causes plant death, and several antioxidative systems that scavenge ROS.

The ROS, which comprise hydroxyl free radicals, superoxide radicals, hydrogen peroxide, and singlet oxygen, lead to DNA mutations, lipid peroxidation, protein denaturation, and disturbance of cellular homeostasis. Plants are protected from the harmful effects of ROS by a sophisticated antioxidant system made up of enzymatic antioxidants and non-enzymatic compounds [2]. The enzyme MDHAR [45], DHAR [44], SOD [44], CAT [45], GR [44], APX [44, 46], GPX [25] and ascorbate-glutathione cycle enzyme are examples of enzymatic antioxidants shown in **Figure 1**. Ascorbate (AsA) [44] and glutathione (GSH) [15, 46] are examples of non-enzymatic antioxidants found in cells shown in **Figure 2**. Increased drought stress levels cause both enzymatic and non-enzymatic antioxidant activity in rice to rise. A strategy against oxidative stress and an improvement in drought tolerance in rice can be achieved by increasing the expression of the antioxidant system [15, 44]. These antioxidant defence enzymes' tendency to be more active reveals their protective role in preventing oxidative damage brought on by drought stress [47].

2.3.3 The role of polyamines in drought stress

Rice responds to drought stress by producing small, positively charged molecules called polyamines (PAs) [12, 47]. Plants contain the PAs putrescine (Put), spermidine (Spd), and spermine (Spm). It can interact with several signalling networks and control homeostasis, membrane stabilisation, and osmotic potential and ionic balance. Increased photosynthetic capability, less water loss, and improved osmotic detoxification and adjustment are all directly related to the PA content rise during drought stress [2, 4]. In response to stress, rice produces significantly more putrescine, which encourages the production of spermidine and spermine and eventually protects the plants from dehydration, according to a recent study on the crop [12, 47].

2.3.4 The role of phytohormones under drought stress

Phytohormones are known to play vital roles in regulating various phenomenons in plants to acclimatise to varying drought environment. Abscisic acid (ABA), cytokinins (CK), Jasmonic acid (JA), ethylene (ET), auxins (IAA), gibberellins (GAs) and other major plant hormones are significant in drought response. However, these hormones are usually cross talk with each other to increase the survival of plants in drought condition [2, 23, 48]. Drought stress is experienced as a hydraulic pull brought on by a pressure gradient between the soil and plants as a result of soil drying. The concentration of the signal hormones ABA shifts in response to the perception of a hydraulic force [48, 49]. While other hormones like CKs may be decreased by down-regulating gene expression, degrading via oxidase enzyme activity, or due to stress damage, ABA concentration normally increases in order to communicate the signals associated with drought stress [48, 49]. Since hormone concentration can function independently to confer a signal or it can act in concert with other hormones and/or other signals, these changes are dynamic and complex. Furthermore, the endogenous concentration of a given hormone may be influenced by the duration and severity of drought stress and may differ in different plant organs. One illustration of how hormones interact with one another is the indirect role played by ABA in water stress signalling by suppressing the production of ET [49, 50]. In response to drought, both ABA-dependent and ABA-independent signalling pathways are activated, and a fast buildup of ABA has been linked to improved drought tolerance [2, 4, 49]. In studies of the highly drought tolerant resurrection plants (*Craterostigma wilmsii*), ABA concentrations were shown to be the most highly affected hormone in response to drought stress [50]. ABA and other hormonal signalling pathways lead to major changes in plant growth, defence responses, and major drought tolerance mechanisms [4].

3. The conventional breeding approaches for improvement of drought tolerance in rice

Normally in conventional breeding methods, grain yield is used as a selection criterion for superior cultivars in drought-prone areas; however this has been demonstrated to be ineffective due to poor heritability and the large impact of genotype by environment interaction [30, 43]. Selection in traditional breeding has steadily moved away from other criteria in favour of physiological qualities since they grow more quickly and depend on genetic variation [43, 51, 52]. However, the main objective of crop breeding is to create high yielding varieties under well water circumstances, and good yielding varieties may continue to produce a high to moderate yield when there is a drought [53]. To put it simply, traditional breeding techniques are crucial for the preservation of germplasm, sexually different parent hybridization, and the emergence of novel genetic characteristics. Using traditional breeding methods, the International Rice Research Institute (IRRI) and the Indian Institute of Rice Research (IIRR), Hyderabad, Telangana, India, have created a wide variety of elite cultivars that are resilient to many diseases and abiotic stresses over the past three decades [2, 53]. In recent years, backcrossing, forced mutation, and pedigree selection have supplanted other traditional breeding techniques as the main ones.

3.1 Pedigree selection

One of the most traditional and popular breeding techniques in rice development is pedigree selection. In particular, if the trait is controlled by important genes, this approach is very suitable for building resistance in rice. The ability to combine numerous genes affecting biotic and abiotic processes is one of the main benefits of pedigree selection [53]. The primary drawback of pedigree selection is that it takes a lot of time and necessitates evaluating numerous lines repeatedly over planting seasons while maintaining a record of the selection criteria. This approach is not appropriate for traits that are influenced by several genes; in this situation, the diallel mating design will be appropriate for selection [53]. Recurrent selection is typically preferred by plant breeders over pedigree selection in the majority of self-pollinating crops, including rice [54]. **Figure 3** showed the general selection process for the development of drought-tolerant rice.

3.2 Recurrent selection

In order to increase favourable allele frequencies while preserving genetic diversity, recurrent selection is utilised in varietal improvement. It offers more accurate genetic gains, quicker and more defined breeding cycles, and the creation of extremely diversified breeding lines. This approach, which outperforms the pedigree selection method, has been extensively explored in rice [55].

3.3 Backcross breeding

In order to reduce the genome of the donor parent and consequently increase high recovery of the recipient parent, the backcrossing technique is frequently employed in rice breeding to introduce desirable or target genes controlling a certain trait from donor parent to recipient parent [56]. This method offers a precise and accurate method for creating numerous superior breeding lines. Backcrossing techniques have facilitated the creation of rice cultivars that can withstand drought [56–59].

3.4 Induced mutation

Since induced mutation has been shown to be effective in the development of improved agronomic traits like an increase in grain yield [57–59], resistance to pests and diseases, and improvement of physical grain quality, it is used to supplement conventional breeding methods [58]. The creation of gene alleles that are not found in nature is the main benefit of induced mutation, according to [57] who summarised the use of

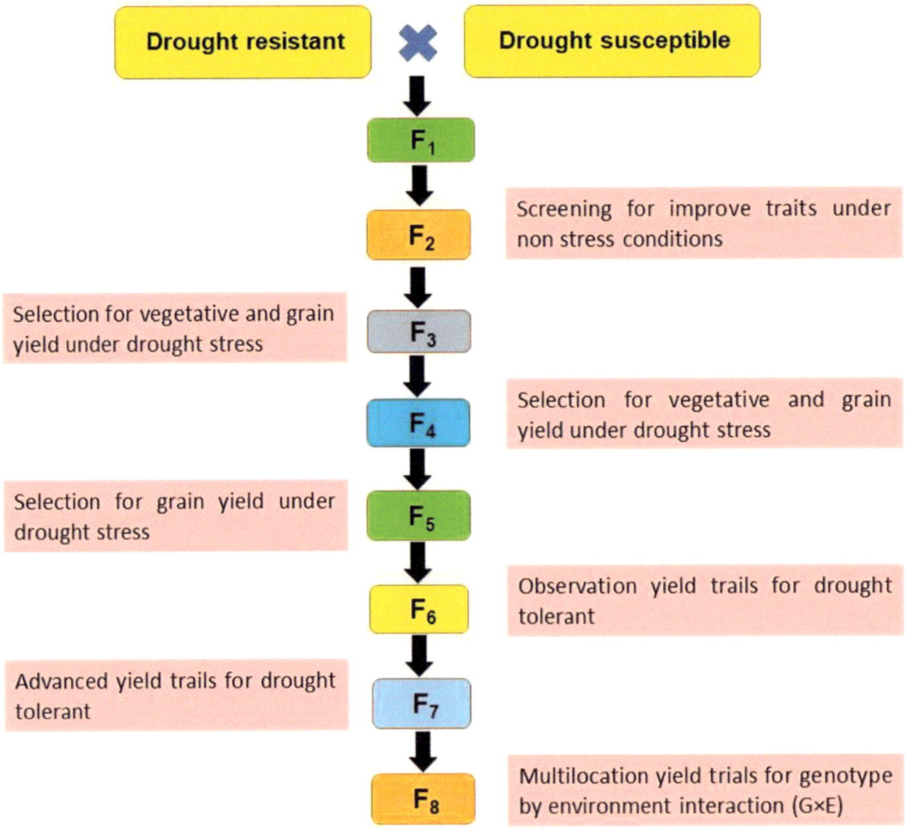

Figure 3.
Modified method for conventional yield trail in rice.

induced mutation with numerous success stories on innovative rice varieties created through induced mutation. Induced mutation is a technique used in plant mutation breeding to create new varieties. Manawthukha rice was exposed to ^{300} Gy of gamma radiation from a ^{60}Co source in Myanmar to test the variety for its ability to withstand drought by withholding irrigation from 90 days after transplant until harvest. Two mutant lines, MK-D-2 and MK-D-3, were determined to be drought resistant after six generations of evaluation and selection by utilising physiological screening procedures [60]. Similar to this, 11 lines with drought-tolerant traits were chosen from an Iranian rice landrace called "Tarom Mahalli" after being exposed to gamma radiation at an optimal dose of ^{230} Gy [61]. Induced mutation allowed scientists in Indonesia to create a super green rice mutant that is drought resilient, high producing, and water efficient [62]. Two better lines, MR219-9 and MR219-4, with high production potential and drought tolerance were developed from the common MR219 rice variety in Malaysia [63].

At final, before initiation of new molecular techniques, our understanding of how plants respond to drought at the molecular and whole-plant level has rapidly expanded. There are hundreds of genes that are expressed under drought stress, and some of them have been cloned. The development of drought tolerance often uses a variety of techniques, such as transgenics and gene expression patterns. Proteomics, genome-wide association, stable isotopes, and fluorescence or thermal imaging's are a few recent techniques that have helped close the genotype–phenotype gap. Rice has developed resilience to drought thanks to genetic engineering and molecular technology, which are the main tools in biotechnological procedures. The most effective and reliable methods to lessen the effects of drought are, in general, the development of genetic resistance.

4. The molecular basis for improvement of drought tolerance in rice

Environmental drought stimuli are detected by yet-to-be-fully-depicted sensors on the membranes, and the signals are then transmitted down through various signal transduction pathways, resulting in the outflow of drought responsive qualities with appropriate gene functions and tolerance to the drought [7, 64]. The phenomenon of drought is complex [22, 65]. So, when it comes to drought tolerance, hybridization and selection techniques cannot provide precise findings. However, the use of DNA markers in molecular investigations can affix the process while still producing precise results. Additionally, by sorting through a large collection of germplasm for drought-tolerant varieties, these molecular markers can be a godsend for further crop improvement. Numerous efforts have been made to identify some qualitative trait loci (QTLs) associated with different attributes [22, 30, 66]. In conclusion, several QTLs for rice drought tolerance have been found **Table 1**. There have not been many researches on grain yield, though. The vast majority of QTLs in rice have been discovered based on a variety of significant features, such as root and shoot responses, osmotic adjustment, hormonal responses, photosynthesis, and whole plant responses to drought tolerance. DNA studies based on marker-based phenotyping are the main methods used to identify genes involved in rice drought resistance. Despite the advancements, only a few numbers of characteristics have realistically been recognised as having drought resistance capacity [66, 86]. To create transgenic crops with improved resistance to drought stress, it is imperative to identify the candidate genes responsible for plant tolerance under various abiotic stresses [87]. Using genetic engineering (*Agrobacterium tumefaciens or gene gun*) and hybridization with marker-assisted selection, the gene governing drought tolerance can then be introduced into the genetic background of any suitable cultivar. In this way, molecular breeding can improve crop types and yield varieties, resulting in prolific harvests with high agronomic validity and safety.

4.1 QTLs associated with rice drought tolerance

The plant genome has a number of genes known as QTLs that have extremely precise quantitative properties. **Table 2** displays many QTLs connected to various agronomic traits under drought. Earlier molecular genetic studies [52, 66, 97, 98] identified a large number of QTLs linked to various physiological and biochemical traits, but were unable to identify genes that regulate these traits due to poor mapping resolution and weak phenotypic effect [17, 18]. Finding these QTLs associated with selected characteristics aids plant stress screening programmes [99]. Selection of drought-tolerant rice cultivars has made significant use of many QTLs connected to many physiological and growth parameters under drought [52, 96, 99, 100]. Additionally, the classification of QTLs at various stages of rice growth is investigated [52, 96, 99, 100]. Considering yield to be a definite point, continuing research institutes worldwide focus primarily on mapping QTLs for grain yield of rice under drought stress [15]. Thus, specific QTLs for drought tolerance might be found and exploited to create drought-tolerant rice cultivars.

Trait	Pedigree	Marker	Mapping population	No. of QTL	References
Seedling drought resistance	Indica × Azucena	RFLP, AFLP, SSR	RIL	7	[67]
Cellular membrane stability	IR62266 x CT9993	RFLP, AFLP, SSR	DH	9	[68]
Leaf water relations and rolling	Azucena × Bala	RFLP, AFLP, SSR	RIL	13	[69]
Seed fertility, spikelet/ panicle, grain yield	Teqing x Lemont	SNP	IL	5	[70]
Root number, thickness, and length	IR58821 × IR52561	AFLP & RFLP	RIL	28	[71]
Root architecture and distribution	IR64 x Azucena	RFLP	DH	39	[72]
Root traits and penetration index	IR1552 × Azucena	SSR	RIL	23	[73]
Deep roots	3 populations	SSR, SNP	RIL	6	[74]
Root penetration, root and tiller number	CO39 × Moroberekan	RFLP	RIL	39	[75]
Root-penetration	Azucena × Bala	RFLP, AFLP	RIL	18	[76]
Grain yield under drought	Two population	SSR	BS	4	[77]
Grain yield in aerobic environments	Three populations	SSR	BS	1	[78]
Yield traits at the reproductive stage	IR64 × Cabacu	SNP	RIL	1	[79]
Yield under stress at reproductive stage	swarna x WAB	SSR	BIL	1	[80]
Heritability for grain yield	CT9993 × IR62266	AFLP	DH	1	[81]
Grain yield under severe lowland drought	R77298 x Sabitri	SSR	BC_1	1	[82]
Yield at reproductive stage over environments	Two populations	SSR	BSA	2	[83]
Morphological and physiological traits	IR64 × Azucena	RFLP	DH	15	[84]
Osmotic adjustment and Dehydration tolerance	CO39 × Moroberekan	RFLP	RIL	1	[85]

Table 1.
QTL for yield and yield contributing traits responses under drought stress conditions in rice.

The majority of the QTLs for drought tolerance in rice that have been discovered so far come from non-elite genotypes. The rice plants' QTL *qDTY1.1* is widely applied

Traits	QTL	Reference
Grain Yield	*qDTY1.1, qDTY3.2, qDTY10.1*	[77, 88]
	qDTY1.2, qDTY1.3	[89]
	qDTY2.1	[78]
	qDTY2.2, qDTY2.3	[83]
	qDTHI2.3	[34]
	qDTY3.1, qDTY6.1, qDTY6.2	[10]
	qDTY4.1, qDTY9.1, qDTY10.2	[90]
	qDTY9.1A	[10]
	qDTY12.1	[91]
	qPDL1.2, qPNF3.1	[92]
Leaf rolling	*qlr8.1*	[92]
	qLR9.1	[66]
	qDLR8.1	[10]
Leaf drying	*qLD9.1, qLD12.1*	[66]
Harvest index	*qHI9.1*	[66]
	qSf6	[93]
	qPNF3.1	[91]
Spikelet fertility	*qSF9.1*	[66]
Panicle number	*qgy3.1*	[94]
Plant height	*qPH1.1*	[79]
Flowering day	*qHGW2.2*	[91]
Panicle number, grain weight	*qGy7*	[95]
Panicle length	*qPL-9*	[95]
Grain number	*qDTY8.1*	[77]
Relative water content	*qRWC9.1*	[66]
Transpiration	*qDTR8*	[96]
Total dry matter yield	*qHGW1*	[91]
Days to heading, grain/panicle	*qPSS8.1*	[79]

Table 2.
QTLs associated with drought tolerance in rice for governing yield and yield contributing traits.

as a yield characteristic while they are under drought stress [69]. **Table 2** lists several significant QTLs that have been discovered in various rice lines, including *qDTY2.1* [52], *qDTY2.2* [52], *qDTHI2.3* [101], *qDTY3.1* and *qDTY6.1* [96], *qDTR8* [100], There are also other SSR markers associated with these QTLs [101]. Therefore, it would be beneficial to use these markers for molecularly screening new rice genotypes for drought tolerance. This would allow for quick and accurate profiling of the rice lines. During the reproductive stage of rice studied the genetic mapping of morpho-physiological traits related to drought tolerance. They reported five QTLs, including *qLR9.1, qLD9.1, qHI9.1, qSF9.1, and qRWC9.1,* which control, respectively, leaf rolling, leaf drying, harvest index, spikelet fertility, and relative water content in rice [66].

4.2 Rice drought tolerance via transgenic/genetic engineering and genetic methods

The production of several protein classes, such as transcriptional factors, molecular chaperones, enzymes, and other functional proteins, by plants has allowed them to create dependable routes or signalling chains for stress [101]. These proteins increase the ability of plants to withstand or fight drought. In reality, these genes (regulatory elements and proteins) have been discovered using various genomic techniques in numbers of hundreds or even thousands. As shown in **Table 3**, these genes have been integrated into the rice genome to investigate their impact on drought improvement by overexpression or suppression. In rice, many transcription factors that are encoded by *WRKY* genes regulate various biological processes. In plants, zinc finger proteins are widely distributed, especially those that control stress responses. Both

Gene	Function	Reference
DRO1	Induces root elongation and deeper rooting	[102]
OsDREB1F	Maintains ABA-dependent signalling pathway	[103]
OsDREB2B	Root length and number of root increment	[103]
CYP735A	Maintains cytokinin level	[104]
OsNAC5	Enhances root diameter and grain yield	[105]
SNAC1	Enhances spikelet fertility	[16]
OsbZIP23	Increases grain yield	[106]
AP37	Enhances seed filling and grain weight	[107]
OsbZIP46	Increases grain yield	[108]
OsbZIP71	Enhances seed setting	[109]
EcNAC67	Increases relative water content, delays leaf rolling, higher root and shoot mass	[110]
DsM1	Helps in reactive oxygen species scavenging, maintains drought tolerance at the seedling stage	[111]
OsPYL/RCAR5	Induces stomatal closure, regulates leaf fresh weight	[35]
OsWRKY47	Relatively low yield reduction	[112]
AtDREB1A	Osmolyte accumulation, chlorophyll maintenance, higher relative water content and reduced ion leakage	[111]
TlOsm	Maintains growth, retains higher water content and membrane integrity and improves survival rate	[113]
OsMIOX	Higher reactive oxygen species scavenging enzyme activity and proline content	[114]
Coda	Better yield, higher photosystem II activity, increased detoxification of reactive oxygen species	[115]
OsTPS1	Higher trehalose and proline accumulation	[49]
OsCPK9	Increases drought tolerance through enhanced stomatal closure and better osmoregulation in transgenics	[111]
OsNAC10	Increases tolerance to drought at vegetative stage, enlarges roots and improves grain yield	[105]

Table 3.
Genes associated with different mechanisms of drought tolerance in rice.

monocotyledons and dicotyledons have the *WRKY* genes, which are widely dispersed in plants. Numerous *WRKY* genes have regulatory functions that can be positive or negative in how plants react to various abiotic stressors [4, 116].

By decreasing stomata density and enhancing stomata closure, rice zinc-finger protein (dst mutant) demonstrated increased drought and salt tolerance. However, DST non-mutants alter H2O2 homeostasis, which has an adverse effect on stomata closure [116]. By improving drought tolerance, overexpression of the zinc finger protein *OsZFP252* demonstrated 74–79% greater chances of survival. Additionally, proline and soluble sugar buildup are increased [127].

About 5000 genes are up regulated and 6000 genes are down regulated in rice after drought stress exposure [18, 128]. **Table 4** lists a few of the genes and their associated roles in rice drought tolerance. These genes are divided into three main categories, including those related to membrane transport, signalling, and transcriptional regulation [30, 35]. Under drought stress, they regulate the majority of rice's biochemical, physiological, and molecular systems [8, 64] According to research by [22, 30] numerous genes and transcription factors exhibit differential expression in rice and are employed for transgenic plants under drought stress. The majority of the genes that are controlled by drought have both ABA-dependent and ABA independent regulation mechanisms [8, 64]. *OsJAZ1* reduces drought tolerance in rice by ABA signalling, which coordinates the plant's responses to drought stress-related growth and development [129]. Additionally, some genes are linked to osmoregulation and late embryogenesis abundant (LEA) proteins, which confer water shortage tolerance in rice [30, 64]. Other genes include *OsPYL/RCAR5* and *EcNAC67* delay leaf rolling and cause increased root and shoot mass in rice under situations of water deficiency.

The gene *DRO1* stimulates root elongation and deeper rooting in transgenic rice [35, 102, 110]. The overexpression of *OsDREB2B, CYP735A*, and *OsDREB1F* in rice under drought stress also increases root morphological adaptations [35]. Rice has a *DREB2*-like gene called *OsDRAP1* that confers drought tolerance, according to [116]. It is essential to increase grain yield in rice during droughts, and transgenic methods are used to do this by introducing genes such *OsNAC5, OsLEA3--1* [130], *OsbZIP71* [109], *OsWRKY47* [112], *OsbZIP46* [131]. By examining genes like *EDT1/HDG11, AtDREB1A, OsMIOX,* and *OsTPS1*, as well as osmolytes accumulation, greater antioxidant enzyme activity, and enhanced photosynthesis, it has been found that transgenic rice has improved water use efficiency [120]. Through improved osmoregulation and stomatal closure, *OsCPK9* increases transgenic plants' ability to withstand drought [132]. Under extreme drought and salinity conditions, transgenic plants' survival is improved by overexpressing *OsDREB2A*. [133]. *CDPK7* and *CIPK03/CIPK12* regulate a number of regulatory proteins, signal transduction pathways, and protein kinases in rice [106]. Under drought stress, *OsITPK2* carries decreased levels of inositol triphosphate and ROS homeostasis in rice [117].

The *WRKY* genes respond to drought stress and are crucial for plant development [2]. Under laboratory or glass house circumstances, various genes have been tested for their ability to confer drought resistance in rice using transgenic techniques. Prior to being used in molecular breeding programmes, these genes should be tested in the field. Trehalose, often referred to as tremalose or mycose, is a key component of abiotic stress, including cold and drought. It protects against stress, stabilises proteins against denaturation, and also stores carbs. Trehalose-6-phosphate synthase (TPS) and trehalose-6-phosphate phosphatase (TPP) are the two primary enzymes that catalyse the manufacture of trehalose in plants; buildup of trehalose in rice has been shown to increase drought tolerance. An increase in trehalose, improved

Gene action	Gene	Promoter	Gene transfer methods	Phenotype	References
Genes encoding enzymes that synthesise osmotic and other protectants					
Polyamine synthesis	*ADC*	*Ubi-1*	*Biolistic*	Improved drought tolerance by producing higher levels of putrescine and spermine synthesis	[12]
abscisic acid Metabolism	*CaMV35SP*	*DSM2*	*Agrobacterium*	Oxidative and drought stress resistance and increase of the xanthophylls and non-photochemical quenching	[117]
Amino acid metabolism	*OsOAT*	*Ubi1*	*Agrobacterium*	Improve drought tolerance and increase seed setting	[118]
Reactive oxygen species	*OsSRO1c*	*Ubi1*	*Agrobacterium*	Oxidative stress tolerance and stomata closure regulation	[118]
Protoporphyrinogen oxidase	*PPO*		*Agrobacterium*	Less oxidative damage, and drought tolerance	[119]
Trehalose synthesis	*OsTPS1*	*Actin1*	*Agrobacterium*	Tolerance of rice seedling to drought, cold, and high salinity	[120]
Late embryogenesis abundant (LEA) related genes					
LEA protein gene	*HVA1*	*Actin1*	*Agrobacterium*	Cell membrane stability, higher leaf relative water content and increase in growth under drought stress.	[121]
	HVA1	*Actin1*	*Agrobacterium*	Drought and salinity tolerance	[122]
	OsLEA3-2	*CaMV35S*	*Agrobacterium*	Drought resistance and increase grain per panicle	[117]

Various regulatory genes					
Transcription factor	*AP37*	*OsCc1*	*Agrobacterium*	Improve growth performance under drought stress	[107]
	OsbZIP23	*Ubi1*	*Agrobacterium*	Wide spectrum to salt, drought tolerance and yield improvement	[113]
	OsbZIP72	*CaMV35S*	*Agrobacterium*	Drought resistance and ABA sensitivity	[123]
Harpin protein	*Hrf1*	*CaMV 35S*	*Agrobacterium*	Drought resistance through ABA signalling and antioxidants, and stomata closure regulation	[124]
Jasmonate and ethylene-responsive factor 1	*JERF1*	*CaMV35S*	*Agrobacterium*	Drought resistance	[124]
Ethylene-responsive factor 1	*TSRF1*	*CaMV35S*	*Agrobacterium*	Enhances the osmotic and drought tolerance	[125]
Stress/zinc finger protein	*OsiSAP8*	*CaMV35S*	*Agrobacterium*	Tolerance to salt, drought and cold stress	[126]

Table 4.
Drought tolerant gene that has been tested in rice transferred through genetic engineering/transgenic methods.

drought tolerance, and decreased photo oxidation in the rice plant under cold and salt stress were seen when a fusion TPP/TPS gene from *Escherichia coli* (otsA and otsB) was introduced into rice [134]. In conclusion, this study indicated that engineering drought-tolerant genes into rice's genetic background is promising, provided that a drought-inducible promoter is employed to achieve successful outcomes.

4.3 Marker-assisted selection (MAS) for rice drought tolerance

To find novel genotypes with desirable drought-tolerant features and related genes/loci, the natural genotypic variation in rice can be studied [135, 136]. Through MAS, these novel genotypes can be used in conventional breeding programmes to create rice varieties that are drought tolerant. Breeding programmes are designed to create high yield lines with enhanced quality metrics and then to introduce the cultivars for agricultural use. Drought tolerant rice genotype breeding has been studied in the past [17, 22, 52, 66, 77], but the success rate has been far below expectations due to the challenge of finding suitable donors with a higher tolerant level as well as the nature of its environment-specific nature.

MAS offers the most accurate, environmentally-friendly, fast and economical method of developing superior rice varieties with a certain degree of resistance or tolerance to drought. The IRRI has been the main site for the majority of the marker-assisted breeding techniques used to create drought-tolerant rice varieties in the past 10 years [22, 137]. Several QTLs for drought tolerance in rice are introduced into top cultivars utilising marker-assisted breeding techniques [17]. In the high-yielding variety IR64, they have successfully incorporated QTLs including *qDTY9.1, qDTY2.2, qDTY10.1,* and *qDTY4.1* using a marker-assisted backcrossing method [17]. With the pyramiding of three QTLs, *qDTY2.2, qDTY3.1,* and *qDTY12.1,* [138] created the elite Malaysian rice cultivar MR219 that is drought-tolerant. Three QTLs were incorporated into the development of the rice variety TDK1 by [52] for high yield during drought (*qDTY3.1, qDTY6.1 and qDTY6.2*). Only as a limitation has drought become more significant, and so far no practical steps have been taken to create rice types that are drought tolerant. A large number of high-yielding cultivars, including Swarna, Samba mahsuri, and IR36, which were previously suggested for cultivation in irrigated regions, have been adopted in the drought breeding effort. Because the aforementioned high yielding varieties cannot withstand repeated droughts, significant loss in rice production is observed when these varieties are cultivated by farmers in rainfed ecosystems during the recurrent drought phase [3]. Therefore, improved special rice varieties with high yields during drought and adaptation to a wide range of unfavourable climatic circumstances in the future require additional focus.

5. Future outlooks

The process of developing drought tolerance in rice is challenging and necessitates a thorough understanding of the different morphological, biochemical, physiological, and molecular characteristics. Despite the impressive advancements made by marker-assisted breeding, there are still a number of major obstacles to molecularly breeding rice for drought tolerance. Additionally, the multigenic regulation and complex nature of drought-tolerant characteristics would be a significant roadblock for ongoing and upcoming research in the field. The complex phenomena of rice crop maintenance during drought conditions are governed by the interactions of a number of factors. Transgenic methods are essential for enhancing rice's agronomic qualities and production characteristics, and they would effectively advance the breeding programme for drought resistance. It is important to understand how these genes react in the presence of drought in the field because several genes have been investigated for their ability to confer drought tolerance in rice under laboratory conditions. Even while substantial basic research is being conducted, we still know relatively little about the mechanism behind the whole-plant stress response. Therefore, we must look into how differentiated cells, tissues, and organs respond to stress and make meaningful connections between the data. In order to improve our understanding of drought tolerance and to promote the genetic improvement of drought-tolerant rice varieties, crop breeding can employ advancements in new technologies related to crop physiology, molecular genetics, and breeding methodologies in an integrated manner.

6. Conclusion

Evaluation for drought resistance is made more difficult by the dynamic and highly variable timing, persistence, and intensity of drought stress under natural settings. Due to their connections to drought stress, abiotic stressors including salt and high temperatures should also be evaluated in conjunction with drought resistance. Many attempts have been made using the numerous QTLs for drought resistance that have been found in rice. High throughput genotyping is now achievable thanks to recent developments in functional genomics, which aid in the identification of key QTL linked to drought tolerance. Therefore, a better knowledge of the genetic basis of drought resistance will be made possible by the successful cloning of these QTL for drought features. The most practical use of drought resistance QTLs, however, is to execute marker-assisted selection based on pyramiding of advantageous QTL alleles to generate drought-resistance in rice utilising recently developing breeding approaches like GWS and MARS. Many genes have been discovered and exploited for enhancing drought resistance via transgenic methods, although the majority of the study was done in glasshouses. The genes that have been shown to be drought tolerance should therefore undergo additional field testing due to the complexity of field settings before being included in the breeding programme. Similar to this, most studies on drought resilience concentrate on the above-ground features, leaving a significant gap for below-ground traits, mostly because phenotyping is challenging. Due to their significant functions in regulating growth and stomata under drought circumstances, root flexibility and architecture should receive enough consideration in studies of drought tolerance.

Acknowledgements

All authors offered equal suggestions on various drafts of the manuscript preparations.

Conflict of interest

The authors declare no conflict of interest.

Author details

Banoth Madhu[1*], Bhimireddy Sukrutha[1], Nunavath Umil Singh[2]
and Govada Venkateswarao[1]

1 Centre for Plant Breeding and Genetics, Tamil Nadu Agricultural University, Coimbatore, Tamil Nadu, India

2 Department of Agronomy, Tamil Nadu Agricultural University, Coimbatore, Tamil Nadu, India

*Address all correspondence to: madhubanoth3596@gmail.com

References

[1] Samal R, Roy PS, Sahoo A, Kar MK, Patra BC, Marndi BC, et al. Morphological and molecular dissection of wild rices from eastern India suggests distinct speciation between *O. rufipogon* and *O. nivara* populations. Scientific Reports. 2018;**8**(1):2773

[2] Panda D, Mishra SS, Behera PK. Drought tolerance in rice: Focus on recent mechanisms and approaches. Rice Science. 2021;**28**(2):119-132

[3] Hao ZC, Singh VP, Xia YL. Seasonal drought prediction: Advances, challenges, and future prospects. Reviews of Geophysics. 2018;**56**(1):108-141

[4] Sahebi M, Hanafi MM, Rafii MY, Mahmud TM, Azizi P, Osman M, et al. Improvement of drought tolerance in rice (*Oryza sativa* L.): Genetics, genomic tools, and the WRKY gene family. BioMed Research International. Oct 2018

[5] Melandri G, AbdElgawad H, Riewe D, Hageman JA, Asard H, Beemster GTS, et al. Biomarkers for grain yield stability in rice under drought stress. Journal of Experimental Botany. 2020;**71**(2):669-683

[6] Turral H, Burke JJ, Faurès JM. Climate Change, Water and Food Security. Rome, Italy: Food and Agriculture Organization of the United Nations; 2011

[7] Oladosu Y, Rafii MY, Samuel C, Fatai A, Magaji U, Kareem I, et al. Drought resistance in rice from conventional to molecular breeding: A review. International Journal of Molecular Sciences. 2019;**20**(14):3519

[8] Gupta A, Rico-Medina A, Caño-Delgado AI. The physiology of plant responses to drought. Science. 2020;**368**:266-269

[9] Myers SS, Smith MR, Sarah G, Golden CD, Bapu V, Mueller ND, etal. Climate change and global food systems: Potential impacts on food security and undernutrition. The Annual Review of Public Health. 2017;**38**:259-277

[10] Dixit S, Singh A, Kumar A. Rice breeding for high grain yield under drought: A strategic solution to a complex problem. International Journal of Agronomy. Oct 2014

[11] Chukwu SC, Rafii MY, Ramlee SI, Ismail SI, Hasan MM, Oladosu YA, et al. Bacterial leaf blight resistance in rice: A review of conventional breeding to molecular approach. Molecular Biology Reports. 2019;**46**:1-14

[12] Capell T, Bassie L, Christou P. Modulation of the polyamine biosynthetic pathway in transgenic rice confers tolerance to drought stress. Proceedings of the National Academy of Sciences of the United States of America. 2004;**101**:9909-9914

[13] Polania J, Rao IM, Cajiao C, Grajales M, Rivera M, Velasquez F, et al. Shoot and root traits contribute to drought resistance in recombinant inbred lines of MD 23-24× SEA 5 of common bean. Frontiers in Plant Science. 2017;**8**:296

[14] Nelson GC, Valin H, Sands RD, Havlík P, Ahammad H, Deryng D, et al. Climate change effects on agriculture: Economic responses to biophysical shocks. Proceedings of the National Academy of Sciences of the United States of America. 2014;**111**(9):3274-3279

[15] Pandey V, Shukla A. Acclimation and tolerance strategies of rice under drought stress. Rice Science. 2015;**22**(4):147-161

[16] Hu H, Xiong L. Genetic engineering and breeding of drought-resistant crops. Annual Review of Plant Biology. 2014;**65**:715-741

[17] Singh R, Singh Y, Xalaxo S, Verulkar S, Yadav N, Singh S, et al. From QTL to variety-harnessing the benefits of QTLs for drought, flood and salt tolerance in mega rice varieties of India through a multi-institutional network. Plant Science. 2016;**242**:278-287

[18] Bin Rahman ANM, Zhang JH. Flood and drought tolerance in rice: Opposite but may coexist. Food and Energy Security. 2016;**5**(2):76-88

[19] Torres RO, McNally KL, Cruz CV, Serraj R, Henry A. Screening of rice Genebank germplasm for yield and selection of new drought tolerance donors. Field Crops Research. 2013;**147**:12-22

[20] Sinclair TR. Challenges in breeding for yield increase for drought. Trends in Plant Science. 2011;**16**:289-293

[21] Rollins JA, Habte E, Templer SE, Colby T, Schmidt J, von Korff M. Leaf proteome alterations in the context of physiological and morphological responses to drought and heat stress in barley (*Hordeum vulgare* L.). Journal of Experimental Botany. 2013;**64**(11):3201-3212

[22] Kumar A, Basu S, Ramegowda V, Pereira A. Mechanisms of drought tolerance in rice. In: Sasaki T, editor. Achieving Sustainable Cultivation of Rice. Vol. 11. UK: Burleigh Dodds Science Publishing Limited; 2016. DOI: 10.19103/AS.2106.0003.08

[23] Farooq M, Hussain M, Wahid A, Siddique KHM. Drought stress in plants: An overview. In: Plant Responses to Drought Stress. Berlin, Heidelberg: Springer; 2012. pp. 1-33

[24] Kadam NN, Tamilselvan A, Lawas LMF, Quinones C, Bahuguna RN, Thomson MJ, et al. Genetic control of plasticity in root morphology and anatomy of rice in response to water deficit. Plant Physiology. 2017;**174**(4):2302-2315

[25] Mishra SS, Panda D. Leaf traits and antioxidant defense for drought tolerance during early growth stage in some popular traditional rice landraces from Koraput, India. Rice Science. 2017;**24**(4):207-217

[26] Vibhuti V, Shahi C, Bargali K, Bargali SS. Seed germination and seedling growth parameters of rice (*Oryza sativa* L.) varieties as affected by salt and water stress. Indian Journal of Agricultural Sciences. 2015;**85**(1):102-108

[27] Sarvestani ZT, Pirdashti H, Sanavy SAMM, Balouchi H. Study of water stress effects in different growth stages on yield and yield components of different rice (*Oryza sativa* L.) cultivars. Pakistan Journal of Biological Sciences. 2008;**11**(10):1303-1309

[28] Hussain HA, Hussain S, Khaliq A, Ashraf U, Anjum SA, Men SN, et al. Chilling and drought stresses in crop plants: Implications, cross talk, and potential management opportunities. Frontiers in Plant Science. 2018;**9**:393

[29] Zhu R, Wu FY, Zhou S, Hu T, Huang J, Gao Y. Cumulative effects of drought-flood abrupt alternation on the photosynthetic characteristics of rice. Environmental and Experimental Botany. 2020;**169**:103901

[30] Upadhyaya H, Panda SK. Drought stress responses and its management in rice. In: Hasanuzzaman M, Fujita M, Nahar K, Biswas JK, editors. Advances in Rice Research for Abiotic Stress Tolerance. UK: Elsevier; 2019. pp. 177-200

[31] Anjum SA, Ashraf U, Zohaib A, Tanveer M, Naeem M, Ali I, et al. Growth and development responses of crop plants under drought stress: A review. Zemdirbyste. 2017;**104**:267-276

[32] Comas LH, Becker SR, Cruz VMV, Byrne PF, Dierig DA. Root traits contributing to plant productivity under drought. Frontiers in Plant Science. 2013;**4**:442

[33] Manivannan P, Jaleel CA, Sankar B, Kishorekumar A, Somasundaram R, Lakshmanan GMA, et al. Growth, biochemical modifications and proline metabolism in *Helianthus annuus* L. as induced by drought stress. Colloids and Surfaces B: Biointerfaces. 2007;**59**(2):141-149

[34] Mishra SS, Behera PK, Panda D. Genotypic variability for drought tolerance-related morpho-physiological traits among indigenous rice landraces of Jeypore tract of Odisha, India. Journal of Crop Improvement. 2019;**33**:254-278

[35] Kim Y, Chung YS, Lee E, Tripathi P, Heo S, Kim KH. Root response to drought stress in rice (*Oryza sativa* L.). International Journal of Molecular Sciences. 2020;**21**(4):1513

[36] Farooq M, Kobayashi N, Wahid A, Ito O, Basra SMA. Strategies for producing more rice with less water. Advances in Agronomy. 2009;**101**:351-388

[37] Ozga JA, Kaur H, Savada RP, Reinecke DM. Hormonal regulation of reproductive growth under normal and heat-stress conditions in legume and other model crop species. Journal of Experimental Botany. 2016;**68**:1885-1894

[38] Fahad S, Bajwa AA, Nazir U, Anjum SA, Farooq A, Zohaib A, et al. Crop production under drought and heat stress: Plant responses and management options. Frontiers in Plant Science. 2017;**8**:1147

[39] Hayat S, Hayat Q, Alyemeni MN, Wani AS, Pichtel J, Ahmad A. Role of proline under changing environments: A review. Plant Signaling & Behavior. 2012;**7**(11):1456-1466

[40] Keunen E, Peshev D, Vangronsveld J, van den Ende W, Cuypers A. Plant sugars are crucial players in the oxidative challenge during abiotic stress: Extending the traditional concept. Plant, Cell & Environment. 2013;**36**(7):1242-1255

[41] Krasensky J, Jonak C. Drought, salt, and temperature stressinduced metabolic rearrangements and regulatory networks. Journal of Experimental Botany. 2012;**63**(4):1593-1608

[42] Liang X, Zhang L, Natarajan SK, Becker DF. Proline mechanisms of stress survival. Antioxidants & Redox Signaling. 2013;**19**:998-1011

[43] Chourasia KN. Resistance tolerance mechanism under water deficit (drought) condition in plants. International Journal of Current Microbiology and Applied Sciences. 2017;**6**:66-78

[44] Gill SS, Tuteja N. Reactive oxygen species and antioxidant machinery in abiotic stress tolerance in crop plants. Plant Physiology and Biochemistry. 2010;**48**(12):909-930

[45] Qureshi MK, Munir S, Shahzad AN, Rasul S, Nouman W, Aslam K. Role

of reactive oxygen species and contribution of new players in defense mechanism under drought stress in rice. International Journal of Agriculture and Biology. 2018;**20**(6):1339-1352

[46] Bhattacharjee S, Dey N. Redox metabolic and molecular parameters for screening drought tolerant indigenous aromatic rice cultivars. Physiology and Molecular Biology of Plants. 2018;**24**(1):7-23

[47] Lum MS, Hanafi MM, Rafii YM, Akmar ASN. Effect of drought stress on growth, proline and antioxidant enzyme activities of upland rice. Journal of Animal and Plant Sciences. 2014;**24**:1487-1493

[48] Raghavendra AS, Gonugunta VK, Christmann A, Grill E. ABA perception and signalling. Trends in Plant Science. 2010;**15**:395-401

[49] Li J, Wu Y, Xie Q, Gong Z. Hormone Metabolism and Signaling in Plants. Cambridge, MA, USA: Academic Press; 2017. pp. 161-202

[50] Luo LJ. Breeding for water-saving and drought-resistance rice (WDR) in China. Journal of Experimental Botany. 2010;**61**:3509-3517

[51] Atkinson NJ, Urwin PE. The interaction of plant biotic and abiotic stresses: From genes to the field. Journal of Experimental Botany. 2012;**63**:3523-3543

[52] Dixit S, Singh A, Cruz MTS, Maturan PT, Amante M, Kumar A. Multiple major QTL lead to stable yield performance of rice cultivars across varying drought intensities. BMC Genetics. 2014:15-16

[53] Khush GS. IRRI breeding program and its worldwide impact on increasing rice production. In: Gene Manipulation in Plant Improvement. Boston, MA, USA: Springer; 1984. pp. 61-94

[54] Miah G, Rafii MY, Ismail MR, Puteh AB, Rahim HA, Asfaliza R, et al. Blast resistance in rice: A review of conventional breeding to molecular approaches. Molecular Biology Reports. 2013;**40**:2369-2388

[55] Pang Y, Chen K, Wang X, Xu J, Ali J, Li Z. Recurrent selection breeding by dominant male sterility for multiple abiotic stresses tolerant rice cultivars. Euphytica. 2017;**213**:268

[56] Lafitte HR, Li ZK, Vijayakumar CHM, Gao YM, Shi Y, Xu JL, et al. Improvement of rice drought tolerance through backcross breeding: Evaluation of donors and selection in drought nurseries. Field Crops Research. 2006;**97**:77-86

[57] Oladosu Y, Rafii MY, Abdullah N, Abdul Malek M, Rahim HA, Hussin G, et al. Genetic variability and selection criteria in rice mutant lines as revealed by quantitative traits. The Scientific World Journal. Oct 2014

[58] Oladosu Y, Rafii MY, Abdullah N, AbdulMalek M, Rahim HA, Hussin G, et al. Genetic variability and diversity of mutant rice revealed by quantitative traits and molecular markers. Agrociencia. 2015;**49**:249-266

[59] Oladosu Y, Rafii MY, Abdullah N, Hussin G, Ramli A, Rahim HA, et al. Principle and application of plant mutagenesis in crop improvement: A review. Biotechnology and Biotechnological Equipment. 2016;**30**:1-16

[60] Soe HM, Myat M, Khaing ZL, Nyo NM, Phyu PT. Development of drought tolerant Mutant from Rice var. Manawthukha through mutation

breeding technique using 60Co gamma source. The International Journal of Innovative Research in Science, Engineering and Technology. 2016;**4**:11205-11212

[61] Hallajian MT, Ebadi AA, Mohammadi M, Muminjanov H, Jamali SS, Aghamirzaei M. Integration of mutation and conventional breeding approaches to develop new superior drought-tolerant plants in rice (*Oryza sativa*). Annual Research & Review in Biology. 2014;**4**:1173-1186

[62] Efendi B, Sabaruddin Z, Lukman H. Mutation with gamma raysirradiation to assemble green super rice tolerant to drought stress and high yield rice (*Oryza sativa* L.). International Journal of Advanced Science, Engineering and Technology. 2017;**5**:1-5

[63] Rahim HA, Zarifth SK, Bhuiyan MAR, Narimah MK, Wickneswari R, Abdullah MZ. Evaluation and characterization of advanced rice mutant line of rice (*Oryza sativa*), MR219-4 and MR219-9 under drought condition. Research Development Seminars. 2012;**44**:26-28

[64] Dash PK, Rai R, Rai V, Pasupalak S. Drought induced signaling in rice: Delineating canonical and non-canonical pathways. Frontiers in Chemistry. 2018;**6**:264

[65] Zargar A, Sadiq R, Naser B, Khan F. A review of drought indices. Environmental Reviews. 2011;**19**(1):333-349

[66] Barik SR, Pandit E, Pradhan SK, Mohanty SP, Mohapatara T. Genetic mapping of morpho-physiological traits involved during reproductive stage drought tolerance in rice. PLoS One. 2019;**14**(12):e0214979

[67] Zheng BS, Yang L, Mao CZ, Huang YJ, Wu P. Mapping QTLs for morphological traits under two water supply conditions at the young seedling stage in rice. Plant Science. 2008;*175*:767-776

[68] Tripathy JN, Zhang J, Robin S, Nguyen TT, Nguyen HT. QTLs for cell-membrane stability mapped in rice (*Oryza sativa* L.) under drought stress. Theoretical and Applied Genetics. 2000;**100**:1197-1202

[69] Khowaja FS, Price AH. QTL mapping rolling, stomata conductance and dimension traits of excised leaves in the Bala × Azucena recombinant inbred population of rice. Field Crops Research. 2008;*106*:248-257

[70] Wang Y, Zhang Q, Zheng T, Cui Y, Zhang W, Xu J, et al. Drought-tolerance QTLs commonly detected in two sets of reciprocal introgression lines in rice. Crop & Pasture Science. 2014;*65*:171-184

[71] Ali ML, Pathan MS, Zhang J, Bai G, Sarkarung S, Nguyen HT. Mapping QTLs for root traits in a recombinant inbred population from two indica ecotypes in rice. Theoretical and Applied Genetics. 2000;**101**:756-766

[72] Yadav R, Courtois B, Huang N, McLaren G. Mapping genes controlling root morphology and root distribution in a doubled-haploid population of rice. Theoretical and Applied Genetics. 1997;**94**:619-632

[73] Zheng BS, Yang L, Zhang WP, Mao CZ, Wu YR, Yi KK, et al. Mapping QTLs and candidate genes for rice root traits under different water-supply conditions and comparative analysis across three populations. Theoretical and Applied Genetics. 2003;**107**:1505-1515

[74] Lou Q, Chen L, Mei H, Wei H, Feng F, Wang P, et al. Quantitative trait locus mapping of deep rooting by linkage and association analysis in rice. Journal of Experimental Botany. 2015;**66**:4749-4757

[75] Ray JD, Yu L, McCouch SR, Champoux MC, Wang G, Nguyen HT. Mapping quantitative trait loci associated with root penetration ability in rice (*Oryza sativa* L.). Theoretical and Applied Genetics. 1996;**92**:627-636

[76] Price AH, Steele KA, Moore BJ, Barraclough PP, Clark LJ. A combined RFLP and AFLP linkage map of upland rice (*Oryza sativa* L.) used to identify QTLs for root-penetration ability. Theoretical and Applied Genetics. 2000;**100**:49-56

[77] Vikram P, Swamy BPM, Dixit S, Trinidad J, Cruz MTS, Maturan PC, et al. Linkages and interactions analysis of major effect drought grain yield QTLs in rice. PLoS One. 2016;**11**(3):e0151532

[78] Venuprasad R, Bool ME, Quiatchon L, Atlin GN. A QTL for rice grain yield in aerobic environments with large effects in three genetic backgrounds. Theoretical and Applied Genetics. 2012;**124**:323-332

[79] Trijatmiko KR, Prasetiyono J, Thomson MJ, Cruz CMV, Moeljopawiro S, Pereira A. Meta-analysis of quantitative trait loci for grain yield and component traits under reproductive-stage drought stress in an upland rice population. Molecular Breeding. 2014;**34**:283-295

[80] Saikumar S, Gouda PK, Saiharini A, Varma CMK, Vineesha O, Padmavathi G, et al. Major QTL for enhancing rice grain yield under lowland reproductive drought stress identified using an *O. sativa*/*O. glaberrima* introgression line. Field Crops Research. 2014;**163**:119-131

[81] Kumar R, Venuprasad R, Atlin GN. Genetic analysis of rainfed lowland rice drought tolerance under naturally-occurring stress in eastern India: Heritability and QTL effects. Field Crops Research. 2007;**103**:42-52

[82] Yadav RB, Dixit S, Raman A, Mishra KK, Vikram P, Swamy BM, et al. A QTL for high grain yield under lowland drought in the background of popular rice variety Sabitri from Nepal. Field Crops Research. 2013;**144**:281-287

[83] Palanog AD, Swamy BM, Shamsudin NAA, Dixit S, Hernandez JE, Boromeo TH, et al. Grain yield QTLs with consistent-effect under reproductive-stage drought stress in rice. Field Crops Research. 2014;**161**:46-54

[84] Hemamalini GS, Shashidhar HE, Hittalmani S. Molecular marker assisted tagging of morphological and physiological traits under two contrasting moisture regimes at peak vegetative stage in rice (*Oryza sativa* L.). Euphytica. 2000;**112**:69-78

[85] Lilley JM, Ludlow MM, McCouch SR, O'toole, J. C. Locating QTL for osmotic adjustment and dehydration tolerance in rice. Journal of Experimental Botany. 1996;**47**:1427-1436

[86] Prakash C, Sevanthi AM, Shanmugavadivel PS. Use of QTLs in developing abiotic stress tolerance in rice. In: Hasanuzzaman M, Fujita M, Nahar K, Biswas JK, editors. Advances in Rice Research for Abiotic Stress Tolerance. UK: Elsevier; 2019. pp. 869-893

[87] Gosal SS, Wani SH, Kang MS. Biotechnology and drought tolerance. Journal of Crop Improvement. 2009;**23**:19-54

[88] Ghimire KH, Quiatchon LA, Vikram P, Swamy BPM, Dixit S, Ahmed H, et al. Identification and mapping of a QTL (*qDTY1.1*) with a consistent effect on grain yield under drought. Field Crops Research. 2012;**131**:88-96

[89] Sandhu N, Singh A, Dixit S, Cruz MTS, Maturan PC, Jain RK, et al. Identification and mapping of stable QTL with main and epistasis effect on rice grain yield under upland drought stress. BMC Genetics. 2014;**15**:63

[90] Swamy BPM, Ahmed HU, Henry A, Mauleon R, Dixit S, Vikram P, et al. Genetic, physiological, and gene expression analyses reveal that multiple QTL enhance yield of rice mega-variety IR64 under drought. PLoS One. 2013;**8**(5):e62795

[91] Sellamuthu R, Chandrababu R, Serraj R. Mapping QTLs for reproductive stage drought resistance traits using an advanced backcross population in upland rice. Crop Science. 2015;**55**:1-12

[92] Lin MH, Lin CW, Chen JC, Lin YC, Cheng SY, Liu TH, et al. Tagging rice drought- related QTL with SSR DNA markers. Crop, Environment & Bioinformatics. 2007;**4**(1):65-76

[93] Wang Y, Zang JP, Sun Y, Ali J, Xu JL, Li ZK. Background-independent quantitative trait loci for drought tolerance identified using advanced backcross introgression lines in rice. Crop Science. 2013;**53**(2):430-441

[94] Lanceras JC, Pantuwan G, Jongdee B, Toojinda T. Quantitative trait loci associated with drought tolerance at reproductive stage in rice. Plant Physiology. 2004;**135**(1):384-399

[95] Zhao XQ, Xu JL, Zhao M, Lafitte R, Zhu LH, Fu BY, et al. QTLs affecting morpho-physiological traits related to drought tolerance detected in overlapping introgression lines of rice (*Oryza sativa*). Plant Science. 2008;**174**(6):618-625

[96] Ramchander S, Raveendran M, Robin S. Mapping QTLs for physiological traits associated with drought tolerance in rice (*Oryza sativa* L.). Journal of Investigative Genomics. 2016;**3**(3):56-61

[97] Alexandrov N, Tai SS, Wang WS, Mansueto L, Palis K, Fuentes RR, et al. SNP-seek database of SNPs derived from 3000 rice genomes. Nucleic Acids Research. 2014;**43**(D1):D1023-D1027

[98] Todaka D, Shinozaki K, Yamaguchi-Shinozaki K. Recent advances in the dissection of drought stress regulatory networks and strategies for development of drought-tolerant transgenic rice plants. Frontiers in Plant Science. 2015;**6**:84

[99] Vinod KK, Krishnan SG, Thribhuvan R, Singh AK. Genetics of drought tolerance, mapping QTLs, candidate genes and their utilization in rice improvement. In: Rajpal V, Sehgal D, Kumar A, Raina S, editors. Genomics Assisted Breeding of Crops for Abiotic Stress Tolerance. Cham: Springer; 2019. pp. 145-186

[100] Lin MH, Lin CW, Chen JC, Lin YC, Cheng SY, Liu TH, et al. Tagging rice droughtrelated QTL with SSR DNA markers. Crop, Environment & Bioinformatics. 2007;**4**(1):65-76

[101] Usman MG, Rafii MY, Martini MY, Yusuff OA, Ismail MR, Miah G. Molecular analysis of Hsp70 mechanisms in plants and their function in response to stress. Biotechnology & Genetic Engineering Reviews. 2017;**33**:26-39

[102] Uga Y, Sugimoto K, Ogawa S, Rane J, Ishitani M, Hara N, et al. Control of root system architecture by *DEEPER ROOTING 1* increases rice yield under drought conditions. Nature Genetics. 2013;**45**:1097-1102

[103] Matsukura S, Mizoi J, Yoshida T, Todaka D, Ito Y, Maruyama K, et al. Comprehensive analysis of rice DREB2-type genes that encode transcription factors involved in the expression of abiotic stress-responsive genes. Molecular Genetics and Genomics. 2010;**283**(2):185-196

[104] Maruyama K, Urano K, Yoshiwara K, Morishita Y, Sakurai N, Suzuki H, et al. Integrated analysis of the effects of cold and dehydration on rice metabolites, phytohormones, and gene transcripts. Plant Physiology. 2014;**164**(4):1759-1771

[105] Jeong JS, Kim YS, Redillas MCFR, Jang G, Jung H, Bang SW, et al. OsNAC5 overexpression enlarges root diameter in rice plants leading to enhanced drought tolerance and increased grain yield in the field. Plant Biotechnology Journal. 2013;**11**(1):101-114

[106] Xiang Y, Tang N, Du H, Ye HY, Xiong L. Characterization of OsbZIP23 as a key player of the basic leucine zipper transcription factor family for conferring abscisic acid sensitivity and salinity and drought tolerance in rice. Plant Physiology. 2008;**148**(4):1938-1952

[107] Oh SJ, Kim YS, Kwon CW, Park HK, Jeong JS, Kim JK. Overexpression of the transcription factor AP37 in rice improves grain yield under drought conditions. Plant Physiology. 2009;**150**(3):1368-1379

[108] Tang N, Zhang H, Li XH, Xiao JH, Xiong LZ. Constitutive activation of transcription factor OsbZIP46 improves drought tolerance in rice. Plant Physiology. 2012;**158**(4):1755-1768

[109] Liu CT, Mao BG, Ou SJ, Wang W, Liu LC, Wu YB, et al. OsbZIP71, a *bZIP* transcription factor, confers salinity and drought tolerance in rice. Plant Molecular Biology. 2014;**84**:19-36

[110] RahmanH,RamanathanV,NallathambiJ, Duraialagaraja S, Muthurajan R. Over-expression of a NAC 67 transcription factor from finger millet (*Eleusine coracana* L.) confers tolerance against salinity and drought stress in rice. BMC Biotechnology. 2016;**16**:35

[111] Ning J, Li XH, Hicks LM, Xiong LZ. A Raf-like MAPKKK gene *DSM1* mediates drought resistance through reactive oxygen species scavenging in rice. Plant Physiology. 2010;**152**(2):876-890

[112] Raineri J, Wang SH, Peleg Z, Blumwald E, Chan RL. The rice transcription factor *OsWRKY47* is a positive regulator of the response to water deficit stress. Plant Molecular Biology. 2015;**88**:401-413

[113] Le TTT, Williams B, Mundree SG. An osmotin from the resurrection plant *Tripogon loliiformis* (*TlOsm*) confers tolerance to multiple abiotic stresses in transgenic rice. Physiologia Plantarum. 2018;**162**(1):13-34

[114] Duan JZ, Zhang MH, Zhang HL, Xiong HY, Liu PL, Ali J, et al. *OsMIOX*, a myo-inositol oxygenase gene, improves drought tolerance through scavenging of reactive oxygen species in rice (*Oryza sativa* L.). Plant Science. 2012;**196**:143-151

[115] Kathuria H, Giri J, Nataraja KN, Murata N, Udayakumar M, Tyagi AK. Glycinebetaine-induced water stress tolerance in *codA*-expressing transgenic

indica rice is associated with up-regulation of several stress responsive genes. Plant Biotechnology Journal. 2009;**7**(6):512-526

[116] Huang XY, Chao DY, Gao JP, Zhu MZ, Shi M, Lin HX. A previously unknown zinc finger protein, DST, regulates drought and salt tolerance in rice via stomata aperture control. Genes & Development. 2009;**23**:1805-1817

[117] Du H, Wang N, Cui F, Li X, Xiao J, Xiong L. Characterization of a β-carotene hydroxylase gene DSM2 conferring drought and oxidative stress resistance by increasing xanthophylls and ABA synthesis in rice. Plant Physiology. 2010;**154**:1304-1318

[118] You J, Zong W, Li X, Ning J, Hu H, Li X, et al. The SNAC1-targeted gene *OsSRO1c* modulates stomata closure and oxidative stress tolerance by regulating hydrogen peroxide in rice. Journal of Experimental Botany. 2013;**64**:569-583

[119] Phung T, Jung HI, Park J, Kim JG, Back K, Jung S. Porphyrin biosynthesis control under water stress: Sustained porphyrin status correlates with drought tolerance in transgenic rice. Plant Physiology. 2011;**157**:1746-1764

[120] Li HW, Zang BS, Deng XW, Wang XP. Overexpression of the trehalose-6-phosphate synthase gene *OsTPS1* enhances abiotic stress tolerance in rice. Planta. 2011;**234**:1007-1018

[121] Babu RC, Zhang J, Blum A, Ho THD, Wu R, Nguyen HT. *HVA1*, a LEA gene from barley confers dehydration tolerance in transgenic rice (*Oryza sativa* L.) via cell membrane protection. Plant Science. 2004;**166**:855-862

[122] Rohila JS, Jain RK, Wu R. Genetic improvement of basmati rice for salt and drought tolerance by regulated expression of a barley *HVA1* cDNA. Plant Science. 2002;**163**:525-532

[123] Lu G, Gao C, Zheng X, Han B. Identification of *OsbZIP72* as a positive regulator of ABA response and drought tolerance in rice. Planta. 2009;**229**:605-615

[124] Zhang L, Xiao S, Li W, Feng W, Li J, Wu Z, et al. Overexpression of a harpin-encoding gene *hrf1* in rice enhances drought tolerance. Journal of Experimental Botany. 2011;**62**:4229-4238

[125] Quan R, Hu S, Zhang Z, Zhang H, Zhang Z, Huang R. Overexpression of an ERF transcription factor *TSRF1* improves rice drought tolerance. Plant Biotechnology Journal. 2010;**8**:476-488

[126] Kanneganti V, Gupta AK. Overexpression of *OsiSAP8*, a member of stress associated protein (SAP) gene family of rice confers tolerance to salt, drought and cold stress in transgenic tobacco and rice. Plant Molecular Biology. 2008;**66**:445-462

[127] Xu DQ, Huang J, Guo SQ, Yang X, Bao YM, Tang HJ, et al. Overexpression of a TFIIIA-type zinc finger protein gene ZFP252 enhances drought and salt tolerance in rice (*Oryza sativa* L.). FEBS Letters. 2008;**582**:1037-1043

[128] Joshi R, Wani SH, Singh B, Bohra A, Dar ZA, Lone AA, et al. Transcription factors and plants response to drought stress: Current understanding and future directions. Frontiers in Plant Science. 2016;**7**:1029

[129] Fu J, Wu H, Ma SQ, Xiang DH, Liu RY, Xiong LZ. OsJAZ1 attenuates drought resistance by regulating JA and ABA signaling in rice. Frontiers in Plant Science. 2017;**8**:2108

[130] Xiao BZ, Huang YM, Tang N, Xiong LZ. Over-expression of a LEA gene in rice improves drought resistance under the field conditions. Theoretical and Applied Genetics. 2007;**115**(1):35-46

[131] Jeong JS, Kim YS, Baek KH, Jung H, Ha SH, do Choi Y, et al. Root-specific expression of OsNAC10 improves drought tolerance and grain yield in rice under field drought conditions. Plant Physiology. 2010;**153**(1):185-197

[132] Wei SY, Hu W, Deng XM, Zhang YY, Liu XD, Zhao XD, et al. A rice calcium-dependent protein kinase OsCPK9 positively regulates drought stress tolerance and spikelet fertility. BMC Plant Biology. 2014;**14**:133

[133] Cui M, Zhang WJ, Zhang Q, Xu ZQ, Zhu ZG, Duan FP, et al. Induced over-expression of the transcription factor OsDREB2A improves drought tolerance in rice. Plant Physiology and Biochemistry. 2011;**49**(12):1384-1391

[134] Jang IC, Oh SJ, Seo JS, Choi WB, Song SI, Kim CH, et al. Expression of a bifunctional fusion of the Escherichia coli genes for trehalose-6-phosphate synthase and trehalose-6-phosphate phosphatase in transgenic rice plants increases trehalose accumulation and abiotic stress tolerance without stunting growth. Plant Physiology. 2003;**131**:516-524

[135] Xu R, Yang Y, Qin R, Li H, Qiu C, Li L. Rapid improvement of grain weight via highly efficient CRISPR/Cas9-mediated multiplex genome editing in rice. Journal of Genetics and Genomics. 2016;**43**:529-532

[136] Shim JS, Oh N, Chung PJ, Kim YS, Choi YD, Kim JK. Overexpression of OsNAC14 improves drought tolerance in rice. Frontiers in Plant Science. 2018;**9**:310

[137] Sandhu N, Kumar A. Bridging the rice yield gaps under drought: QTLs genes and their use in breeding programs. Agronomy. 2017;**7**(2):27

[138] Shamsudin NAA, Swamy BPM, Ratnam W, Cruz MTS, Raman A, Kumar A. Marker assisted pyramiding of drought yield QTLs into a popular Malaysian rice cultivar, MR219. BMC Genetics. 2016;**17**(1):30

Chapter 3

Gaseous Losses of Nitrogen from Rice Field: Insights into Balancing Climate Change and Sustainable Rice Production

Jannatul Ferdous, Farah Mahjabin, Mohammad Abdullah al Asif, Israt Jahan Riza and Mohammad Mofizur Rahman Jahangir

Abstract

The world is confronted with one of the most difficult tasks of the twenty-first century, satisfying society's expanding food demands while causing agriculture's environmental impacts. Rice security is the food security for South Asian countries. Rice production requires a large amount of water and fertilizer, especially nitrogenous fertilizer, where urea works as the primary source of nitrogen (N). Different biogeochemical conditions, such as alternate wetting and drying (AWD), intermittent drainage, agroclimatic conditions, oxic-anoxic condition, complete flooded irrigation,. have severe impacts on GHGs emission and nitrogen use efficiency (NUE) from rice fields. For sustainable production, it is a must to mitigate the emissions of GHGs and increase NUE along with cost minimization. But analytically accurate data about these losses are still not quantifiably justified. In this chapter, we will show the proper use of the measured data with suitable results and discussions to recommend the future cultivation system of rice for sustainable production.

Keywords: sustainable rice production, greenhouse gases, alternate wetting and drying (AWD), nitrogen use efficiency (NUE), biogeochemistry

1. Introduction

Nearly half (3.5 billion) of the global population relies on rice (*Oryza sativa*) for sustenance [1]. High-yielding rice varieties (HYV) are widely used in modern agricultural practices to feed the world's teeming population and are accompanied by an increase in the need for chemical fertilizers, particularly nitrogenous fertilizers. A paddy rice field is the center of nitrogen (N input and output. The average N application at rice fields in Japan is 80 kg ha^{-1} season^{-1}, in the USA is 140 kg ha^{-1} season^{-1}, and in Bangladesh is around 100 kg ha^{-1} season^{-1} [1–3] where the global use of different N-fertilizer was 77 Tg N year^{-1} (**Figure 1**). Application of N fertilizer in the

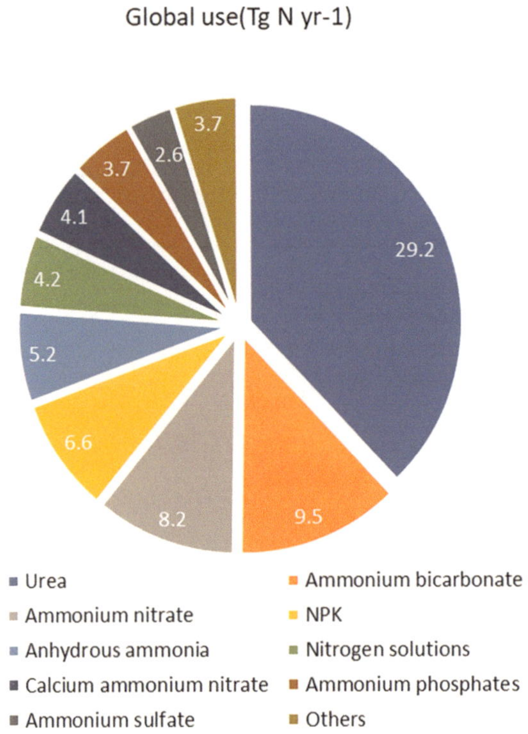

Figure 1.
Globally used different N-fertilizer [4].

paddy rice field is equivalent to more than 60% of the use of the total fertilizers, and it is expected to rise from 107 to 115 million tons (MT) over the 2015/23 period, with an average annual growth rate of 2.4% [5]. However, nitrogen use efficiency (NUE) is only about 30–35%, where the fates of the rest of applied N are leaching, denitrification, nitrification, and ammonia (NH_3) volatilization loss. Rice production generates 1.5% of total anthropogenic greenhouse gas (GHG) emissions globally, but it accounts for 32% of agricultural GHG emissions in Bangladesh [6]. Paddy fields and other agricultural practices, cattle farms, landfills, fossil fuel burning, etc., are major sources of GHGs emissions to the atmosphere.

For rice production, about 3 cm height of standing water is required throughout the rice growing seasons, and about 3500–4000 L of water is required per kg rice grain. Irrigation designs to water save such as alternate wetting and drying (AWD) and dry direct-seeded rice (DSR) planting systems are becoming increasingly important in several rice-growing countries, as it has been found that AWD practice tends to save 38% of water which was needed to be used in irrigation purpose without any reduction in yield [7]. However, in terms of global climatic concerns, AWD has the harsh effect that it emits more N_2O rather than CH_4. Carbon dioxide (CO_2), CH_4, and N_2O are listed as greenhouse gases, but NH_3 is another common gas derived from rice fields and this contributes indirectly to global warming. About 10–60% of the applied nitrogen in rice fields could be lost by NH_3 volatilization, which is the major pathway of N loss, while 5–10% of the applied nitrogen could be lost through denitrification [8, 9]. The amount of N loss as NH_3 in Asia's agriculture is expected to rise from 4.6 Tg N year^{-1} in 1961 to 13.8 Tg N year^{-1} in 2000 [10, 11], and in the next three decades, it will be 18.9 Tg N year^{-1}. The fierce application of N-fertilizer increases the reactive N species in the atmosphere, changes N-cycle, and is subjected to global warming and climate change.

Climate change is mostly driven by the influence of rising temperatures in the earth's atmosphere, where greenhouse gases are responsible for the event. Without regard for the planet's well-being, industrialized and developing countries release GHGs to boost their economy. The greenhouse effect is not only a localized concern as the gases have long-lasting self-life in the atmosphere. The GHGs are not confined to the territory of the producer countries. Beyond the emitting country's border, people must feel the harsh effect of GHGs globally. All countries are not equally responsible for the GHGs effect, but all are the potential to contribute more or less to cause climate change. The most challenging fact in this era is to balance food security and climate change by reducing NH_3, CH_4, and N_2O emissions from rice fields while maintaining or increasing rice production [12]; thus, it is necessary to develop climate-smart technology and techniques.

This part of the chapter reconnoiters the current rice production scenario with the fierce application of N-fertilizers and their contribution to global warming by emitting GHGs. This chapter elaborates on the contribution of rice cultivation to GHG gas emissions and subsequent global warming. Finally, a climate-smart agriculture section will be discussed, which is an effective strategy to minimize GHG emissions and mitigate global warming.

2. Climate change and rice production

The rise in the earth's surface temperature was felt and noticed mostly in the last three decades of the nineteenth century. The earth's surface temperature has been warming at the rate of 1.09°C since the middle of the nineteenth century [13]. Human-induced factors, including GHGs emissions and land use changes, play a vital role in climate change. Where initially, natural factors were also considered for climate change but recently, anthropogenic factors have raised global CO_2 content from 284 ppm (in 1832) to 410 ppm (in 2013) [14], which has now reached 418 ppm and wetland paddy field is a hotspot for CH_4 production [15]. The N_2O concentration increases from 271 ppb (in 1750) to 331 ppb (in 2018) [15]. There has been a 20% increase in N_2O since the industrial era, with an increased rate of 0.95 ppb yearly [15]. These are major GHGs in the atmosphere, responsible for global warming and climate change. Unevenness in climatic conditions is evidence of oscillated temperature changes, precipitation frequency and pattern, and extreme stress events. A global population of about 9 billion people necessitates a 40% increase in rice production by 2030 and a 70% increase in current production by 2050 to feed them. The most densely populated countries in the world, China and India, account for 20% and 28.5% of the total global rice area, respectively [16]. As a result, they contribute significantly to CH_4 emissions [17]. Compared with CO_2, the global warming potential (GWP) of N_2O is 265–298, and the GPW of CH_4 is 34 [15]. Rice at its flowering stage is very susceptible to temperature for pollen viability; hence, global warming is also a concern for food security. Nitrogen fertilizers occupy more than 70% of all the chemical fertilizers in rice fields. However, from winter rice fields, a large portion of applied N is lost to the environment in the form of ammonia (NH_3) through volatilization (about 17%) [2]; this gas has a very short lifespan (a few hours to 12 days) in the atmosphere [15]. Ammonia acts as a potential source of N_2O, chlorofluorocarbon (CFC), and aerosol, and after the break down, it travels as particulate matter [18]. When aerobic rice cultivation is practiced, it will be a great source of N_2O emissions. Nutrient mining, skewed N, P, and K application, poor nutrient use efficiency, and loss of nutrients are key challenges of nutrient use that threaten the economic and environmental sustainability of rice production worldwide [19]. Nutrient management strategy for enhancing rice productivity, especially in an intensive rice-based cropping system, should simultaneously aim to enhance eco-efficiency and sustainability by reducing chemical N fertilizer application. An appropriate rice cultivation practice is needed to increase NUE, preserve natural resources, and make the environment pollution free from GHGs to resist climate change.

3. Why is ammonia a matter of concern for GHGs emissions?

Nitrogen in different forms, for instance, ammonia volatilization, nitrification, and denitrification, all contribute to climate change in varying degrees. Denitrification was discovered to be more significant in Griffith, Australia, whereas volatilization was more significant in Munoz, Philippines [20, 21]. Environmental pollution is also possible due to NH_3 loss as about 17% of applied N lose as NH_3 (**Figure 2**) at the winter rice growing season is possible [2]. Ammonia is the main component of soluble alkaline gas in the atmosphere, enhancing the production of reactive N species as air pollutants (aerosol, nitrous oxide, etc.) causing global warming and significantly affecting local air quality [22]. It can travel long distances before being converted into fine particles [23]. The NH_3 emission increases with fertilizer application rate with

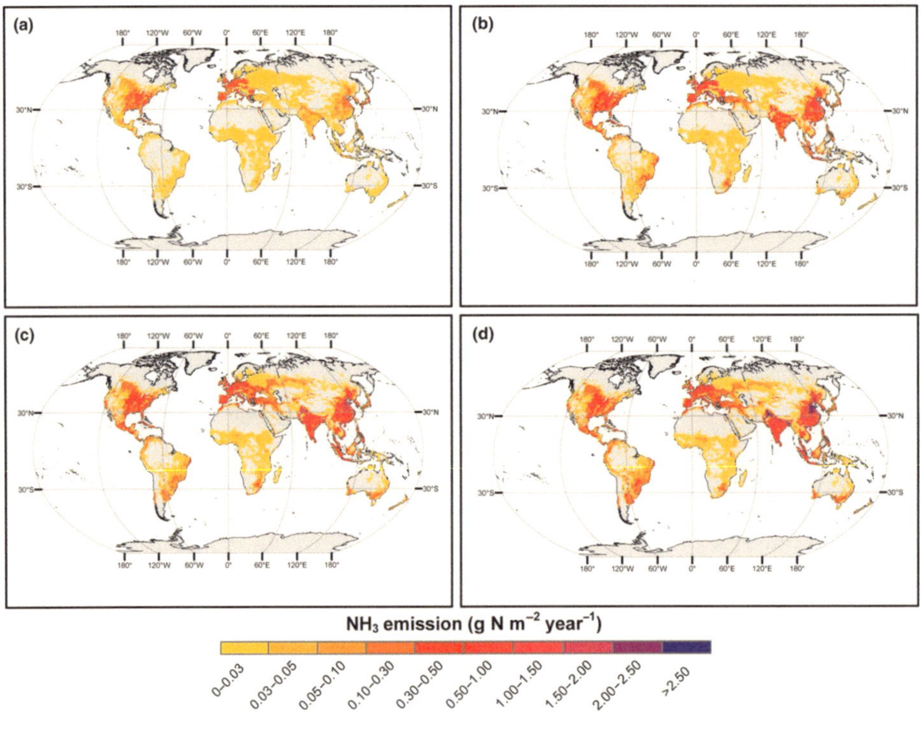

Figure 2.
Simulated ammonia (NH3) emissions in response to application of synthetic nitrogen (N) fertilizer in: (a) the 1960s, (b) the 1980s, (c) the 1990s, and (d) the 2000s [3].

times (**Figure 2**). Consequently, NH_3 deposition negatively affects living organisms as NH_3 reacts with other air pollutants to create tiny particles that can lodge deep in the lungs, causing asthma attacks, bronchitis, and heart attacks [24]. Compared to other GHGs, ammonia has a short atmospheric lifespan (2–10 days) and no direct greenhouse with acids to form salts and return to earth, similar to N-fertilizer [15]. Significant reductions in N_2O and NO emissions from flooded rice soils are an alarming global concern [19]. Therefore, mitigation of NH_3 and N_2O emissions from the cropland system is an urgent demand for environmental and economic protection.

4. Nitrogen (N) transformation pathways from rice fields

4.1 Nitrogen cycle

Nitrogen cycle is a biogeochemical process through which N is converted into many forms, consecutively passing from the atmosphere to the soil to organism and back into the atmosphere. In rice fields, N is applied mostly in the form of urea [25] or urea-containing fertilizer as the principal form of synthetic N. After fertilization, urea immediately undergoes hydrolysis in the presence of standing water, and ammonium bicarbonate is also used in China as N-fertilizer. In rice fields, the water availability facilitates the urease enzyme activity, and the applied urea dissociates into ammonium, bicarbonates, and hydroxyl ions. Among the plant nutrients, N is mostly mobile in nature; thus, total recommended dose of N-fertilizer is divided into two or three equals and applied in two or three splits in rice field through broadcast onto the standing water. First split will be applied on 10–14 days of transplanting (DAT) the maximum tillering stage, at 30–35 DAT second split will be applied, and finally 10–15 days after second split the third split will be applied. The N cycle includes several biological and non-biological processes in a rice field under aerobic and anaerobic conditions. The biological processes are ammonification, mineralization, nitrification, denitrification, N fixation, N assimilatory reduction and microbial synthesis of ammonium and organic N into microbial cells, plant uptake, and conversion of ammonium and nitrate N into plant proteins (**Figure 3**). The non-biological processes are ammonia volatilization, leaching of nitrite and nitrate N to groundwater, ammonium fixation into soil clay minerals, and precipitation of nitrate and ammonium N [26].

4.1.1 Mineralization

Mineralization (or ammonification) of soil N is the term used for the process by which nitrogen in organic compounds is converted by soil microorganisms into ammonium ions (NH_4^+) as following Eq. (1) [27].

$$\text{Complex organic nitrogen} \rightarrow \text{Ammonium} \quad (1)$$

4.1.2 Nitrification

Nitrification is the oxidation of ammonium NH_4^+-N to nitrites and nitrates. The two groups of organisms that are the primary nitrifying bacteria, that is, Nitrosomonas Sp. and Nitrobacter Sp. The general oxidative processes involved can be represented by the following Eqs (2) and (3).

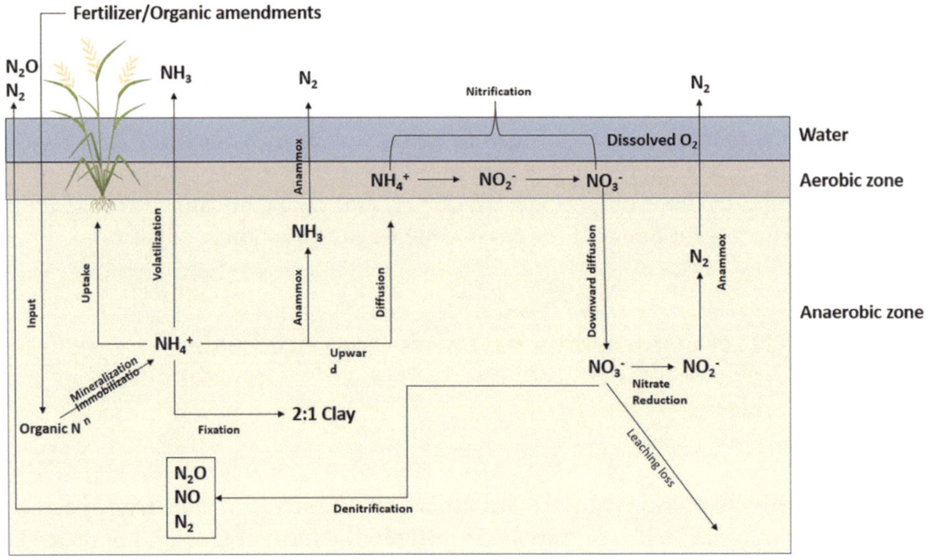

Figure 3.
Nitrogen cycle in rice field.

$$2NH_4^+ + 3O_2 \longrightarrow 2NO_2^- + 4H + 2H_2O \quad (2)$$

$$2NO_2^- + O_2 \longrightarrow 2NO_3^- \quad (3)$$

4.1.3 Denitrification

Denitrification (or nitrate reduction) is a more complex and less understood process than nitrification. Denitrification is a redox process involving nitrogen compounds to obtain energy. There are two processes of nitrate reduction: assimilatory and dissimilatory. When the dissolved oxygen level drops to low levels for the aerobic metabolism of facultative organisms, they can turn to nitrate reduction, presented in the Eq. (4) [28].

$$2NO_3^- \rightarrow 2NO_2^- \rightarrow 2NO \rightarrow 2N_2O \rightarrow N_2 \quad (4)$$

4.1.4 Leaching

Leaching of nitrite or nitrate refers to the removal of nitrite or nitrate from the plant root zone by the movement of water through the soil body. Since nitrite (NO_{2^-}) and nitrate (NO_{3^-}) are negatively charged, they are found to move freely with the water unless soils have a significant anion exchange capacity. It was estimated that 55 Tg of nitrate is leached from agricultural soils every year [29]. Leaching of nitrogen from soil reduces the bioavailability of plants and impacts the environmental quality.

4.1.5 Ammonia volatilization

It is a non-biological process that occurs at the soil surface when ammonia from urea or ammonium-containing fertilizer (urea) is converted to ammonia (NH_3) gas at high pH. Ammonia volatilization occurs when ammonium ions are present in a neutral or in alkaline medium [30]. Due to continuous flooding, the soil pH of rice fields remains 6.5–7 or a little more; thus, ammonia is formed continuously in flooded rice fields through mineralization, which under certain favorable conditions can be lost to the atmosphere as NH_3, presented in Eq. (5). Ammonia can also be produced from waste products of wild animals, cow dung, or compost.

$$NH_4^+ + OH^- \rightarrow NH_3\uparrow + H_2O \tag{5}$$

4.1.6 Anammox

Anammox stands for anaerobic ammonium oxidation, a recent discovery in N-cycle. In the anammox, NH_4^+ is oxidized to nitrite as an electron acceptor and then elemental N_2 by a series of metabolic processes under anaerobic circumstances [8]. This process is carried out by a group of bacteria, that is, annamox, and they are abundant in marine ecosystem so the anammox is a critical issue in the marine ecosystem to understand N-cycle. The anammox bacteria are also found in terrestrial ecosystem. The idea of the N cycle in paddy fields has changed because the paddy field acts as a niche of anammox bacteria by providing favorable conditions (water logging (anaerobic condition) and high NH_{4+} and NO_3^- content) for their living and activity [8]. According to reports, the anammox mechanism is responsible for 4–37% of the nitrogen loss in agricultural soils [31]. The organic content, NO*x* concentration, environmental stability, salinity, and temperature have all been identified as key influencing elements in the ecological dispersion of anammox bacteria and their contribution to N loss in natural environments [32]. The significant loss of N due to anammox occurring in paddy fields is similar to that for NH_3 volatilization (up to 40%), leaching (9–15%), runoff (5–7%), and denitrification (up to 40%) (**Table 1**) [8, 33, 34].

5. Gaseous loss of nitrogen from paddy rice fields

5.1 Nitrous oxide (N_2O) emission

With its current atmospheric concentration of 350 ppb, nitrous oxide (N_2O) is one of the major greenhouse gases, contributing about 5% of the overall greenhouse effect [35]. With a relatively large global warming potential of about 298 times that of carbon dioxide in a 100-year horizon [36], N_2O is one of the main greenhouse gases that cause global warming and ozone depletion [37]. Agricultural activities are responsible for two-thirds of the total anthropogenic nitrous oxide (N_2O) emissions worldwide [38]. Nitrous oxide is emitted into the atmosphere from both natural (about 60%) and anthropogenic sources (approximately 40%), including oceans, soil, biomass burning, fertilizer use, and various industrial processes [39]. Agriculture is a major source of anthropogenic N_2O emissions [40]. During rice production, puddling is operated which normally shuts the water transmission pores resulting in very low water percolation and gaseous exchange between water and air surface. Nitrous oxides are produced from rice fields through nitrification and denitrification processes (**Figure 4**). Emissions of nitric and nitrous oxides are the result of microbial nitrification and denitrification in soils, controlled principally by soil water and mineral N contents, labile organic carbon, and temperature. Nitric oxide is a direct intermediate of both nitrification and denitrification [4]. In submerged soils, nitrification occurs in aerobic sites at the floodwater-soil and root-soil interfaces.

Country/year	2016	2017	2018	2019	2020
Indonesia	5.0973	5.0690	5.2031	5.1137	5.1279
China	6.8563	6.9122	7.0280	7.0562	7.0402
India	3.7902	3.8493	3.9568	4.0577	3.9623
Bangladesh	4.5863	4.6619	4.7257	4.8088	4.7402
Thailand	2.9678	3.0690	3.0380	2.9164	2.9064
Cambodia	3.4443	3.5388	3.5875	3.6721	3.7568
Myanmar	3.8181	3.8218	3.8568	3.7957	3.7711
Pakistan	3.7716	3.8526	2.5629	2.4436	2.5244
Philippines	3.8690	4.0061	3.9718	4.0449	4.0888
Vietnam	5.5738	5.5476	5.8180	5.8371	5.9201

Table 1.
Paddy production (metric tons/ha) of major rice growing countries; (2016–2020); (FAOSTAT).

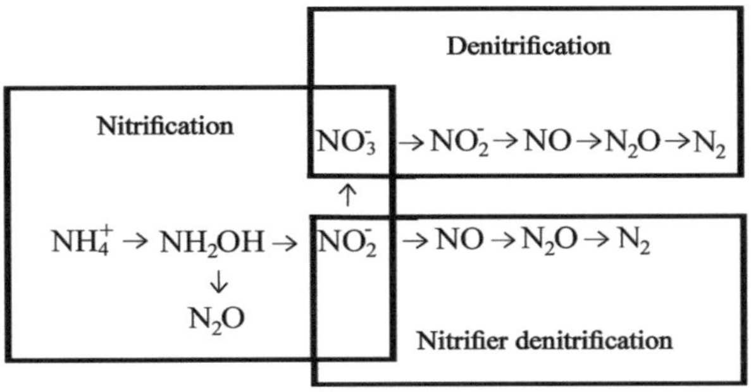

Figure 4.
Principle N transformation pathways leading to the emission of N_2O in soils.

Denitrification occurs upon diffusion of the NO_{3^-} to the anaerobic bulk soil (**Figure 3**). Denitrification is favored over dissimilatory reduction to ($NH_4^+ \rightarrow NO_3^- \rightarrow NO_2^- \rightarrow NH_4^+$) because of the large ratio of available carbon to electron acceptors in submerged soils. Denitrification is likely to proceed completely to N_2 with little accumulation of N_2O because of the very large sink and therefore steep concentration gradient of O_2, and because carbon is less likely to be limiting. However, this will not be the case when submerged soil is drained and air enters, leading to gradients of oxidation from the surfaces of soil cracks toward the anaerobic interiors of soil clods. Now conditions are favorable for the production of nitrous and nitric oxides [4, 41, 42].

The ammonia and nitrate are likely to be converted to N_2O through nitrification and denitrification, but it varies in different cultural practices. Although N_2O emissions from rice fields are substantially smaller than methane emissions in flooded conditions, they have been a long-standing reason for concern [43]. Several researchers have done experiments on the emissions of nitrous oxides from paddy fields. Major rice-producing countries *viz.* China, India, Indonesia, Japan, Philippines, USA contribute to global warming by emitting greenhouse gases like N_2O. **Tables 2** and **3** show the account of how much these countries emit N_2O from their paddy fields and compare different management practices as well.

Country	Location	Type of organic amendment[a]	Amount of organic amendment, kg ha^{-1}	Type of nitrogen fertilizer[b]	Tested mitigation option[c]	Amount of chemical N, kg ha^{-1}	N_2O emission, g ha^{-1}	Reference
Complete flooding condition								
China	Yancheng, Jiangsu	no	0	U		225	2150	Xu et al. [45]
China	Shenyang	no	0	no		0	40	Chen et al. [46]
China	Shenyang	no	0	U		374	60	Chen et al. [46]
China	Shenyang	FYM	37,500	U		374	60	Chen et al. [46]
China	Shenyang	FYM, Azolla	37,500	U		374	60	Chen et al. [46]
India	New Delhi	no	0	no		0	38	Ghosh et al. [47]
India	New Delhi	no	0	U		120	168	Ghosh et al. [47]
India	New Delhi	no	0	U	DCD	120	80	Ghosh et al. [47]
India	New Delhi	no	0	AS	DCD	120	82	Ghosh et al. [47]
India	New Delhi	FYM	6000	U		60	593	Pathak et al. [48]
Indonesia	Bogor, West Java	no	0	no		0	315	Suratno et al. [49]
Indonesia	Bogor, West Java	no	0	U		86	649	Suratno et al. [49]
Japan	Ryugasaki, Ibaraki	Straw	5000	NH_4^+		90	65	
Philippines	Los Banos	no	0	AS		120	97	Abao et al. [50]
USA	Louisiana	no	0	U		0	74	Smith et al. [51]
Mid-season drainage or AWD[d]								
China	Fenqiu	SM	5000	ABC, U		364.5	4416	Cai et al. [52]
China	Yingtan, Jangxi	vetch	33,750	U		276	2810	Xiong et al. [53]
China	Beijing	no	0	ABC		125	2029	Khalil et al. [54]
China	jiangsu	no	0	ABC		191	2389	Zheng et al. [55]
India	New Delhi	no	0	U	DCD	140	142	Kumar et al. [56]

Complete flooding condition								
Country	**Location**	**Type of organic amendment[a]**	**Amount of organic amendment, kg ha^{-1}**	**Type of nitrogen fertilizer[b]**	**Tested mitigation option[c]**	**Amount of chemical N, kg ha^{-1}**	**N_2O emission, g ha^{-1}**	**Reference**
India	New Delhi	no	0	U	TS	140	142	Kumar et al. [56]
India	New Delhi	no	0	no		140	142	Majumdar et al. [57]
India	New Delhi	no	0	no	NEU	140	57	Majumdar et al. [57]
India	New Delhi	FYM	0	no		60	714	Pathak et al. [48]

[a] *Type of organic amendment: FYM, farmyard manure*
[b] *Fertilizer type: AS = ammonium sulfate; U = urea;* NH_4^+ *= complex fertilizer including* NH_4^+.
[c] *Mitigation options: DCD = dicyandiamide. NEU = Nitrogen use efficiency, neem-coated urea; TS = thiosulfate.*
[d] *AWD: Alternate wetting and drying.*
The table shows that N_2O emissions from rice fields are higher in AWD practice than continuous flooding condition. The paddy field can also emit N_2O if remains fallow. The following table shows N_2O emissions from paddy fields during the fallow Period.

Table 2.
N_2O emissions from paddy fields during the cropping season [44].

Country	Location	N_2O Flux, $\mu g\ m^{-2}\ h^{-1}$	N_2O Emission, $g\ N\ ha^{-1}$	Measurement Period in Fallow Season, Days	Reference
China	Shenyang	55.0	3050	231	Chen et al. [46]
China	Shenyang	31.9	888	116	Hou et al. [58]
China	Jiangsu	57.2	2895	229	Zheng et al. [55]
Japan	Tsukuba, Ibaraki	10.6	577	226	Nishimura et al. [59]
Japan	Ryuugasaki, Ibaraki	9.9	606	254	Tsuruta et al. [60]
China	Yingtan, Jangxi	11.0	430	165	Xiong et al. [53]
Philippines	Los Baños	138.7	1198	36	Abao et al. [50]
Philippines	Los Baños	32.4	560	72	Abao et al. [50]

Table 3.
N_2O emissions from Paddy fields during the fallow period [44].

5.2 Ammonia emissions

In the rice field, N-fertilizers are applied in two or three splits. During the first split, the plants are very small and unable to use the applied N fully, resulting in the maximum N loss through volatilization in this period; the standing water in the rice fields favors this process (**Figure 5**) [2]. In the second and third applications, the plants' canopy and root systems will be established, and the N loss through volatilization is less, but the N adsorption by crop increases (**Figure 6**) [2, 4]. Several methods (enclosure method, continuous airflow enclosure method, micrometeorological method, simple low-cost chamber method, wind tunnel method venting method, etc.) are used for ammonia volatilization loss measurements from rice fields (**Table 4**). Often the loss caused at an early stage of plant is about 30–40% loss of applied N and

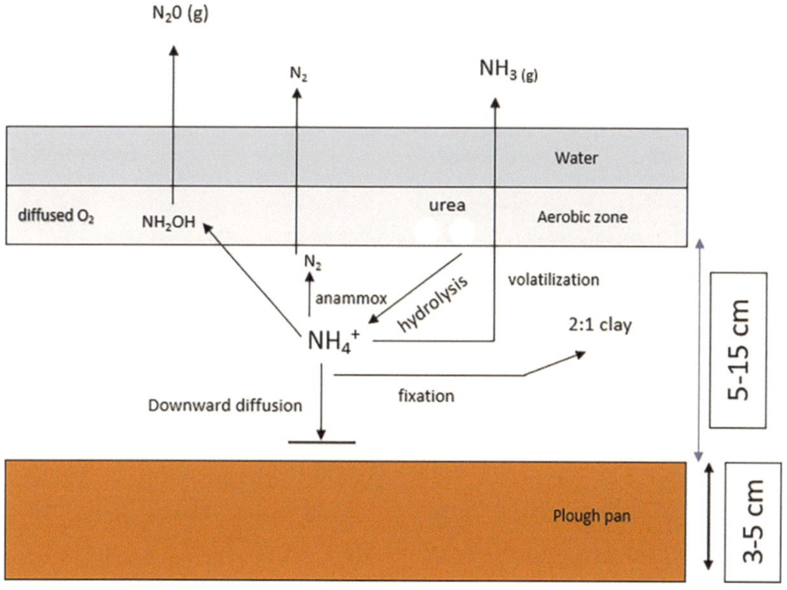

Figure 5.
Role of ammonium in greenhouse effect.

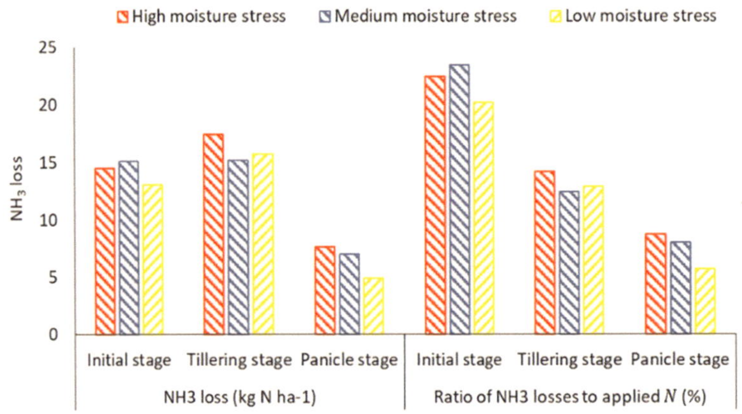

Figure 6.
Ammonia loss at different plant growth stages in presence of water stress [61].

sometimes more than 60% of the applied N, whereas, at later stages, the loss will be half of the earlier one [20, 33]. In rice fields, the loss is responsive to the crop demand and applied N rate. The NH_{4+} ions have a positive relation with ammonia volatilization loss (**Figure 7b**) and N-fertilizer application [65]. In rice field, the NH_{4+} concentration varies from 0 to 1.72 mg/l [66], 0.2–4.5 mg/l [67] and 2–9.2 mg/l [65, 68] due to variation in soil properties (clay particle), soil pH (**Figure 7a**), agriculture operation, climatic condition, hydraulic properties, irrigation, and nitrogen management practices. It becomes more likely that the equilibrium will shift from non-volatile NH_{4+} to volatile NH_3 gas as the pH of the rice field water rises. The ionized ammonia (NH_{4+}) releases 1

Region	Year	Method	NH_3 emission	References
Global	2000	DLEM-Bi-NH3	13.6 ± 0.5	Xu et al. [3]
	1995		12.4 ± 0.3	Xu et al. [3]
	2000	IPCC Tier 1 guideline	7.7	IPCC [62]
	2000	Process-based model	12.0	Riddick et al. [63]
	1995	Constant EF	9.0	Bouwman et al. [64]

Table 4.
Estimates of global NH_3 emissions (expressed in Tg N year^{-1}) based on different approaches [3].

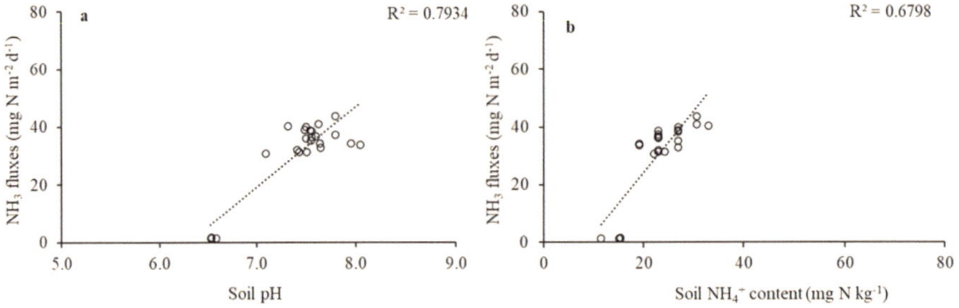

Figure 7.
Relationship between (a) NH_3 flux and soil pH (b) NH_3 flux and NH_4^+.

mole H^+ during subsequent volatilization of non-ionized ammonia (NH_3). The NH_{4^+} is primarily formed in the soil because urease activities are much higher than in the floodwater, so urea moves downward through mass flow and diffusion. The produced ions move between soil and floodwater, and ammonium ion converted into ammonia gets lost in the atmosphere in the volatilization process. Nitrogen loss through NH_3 volatilization can be 20–30 kg ha^{-1} [69] and NH_3 loss can be 46% in rangeland and if the N fertilizer increased 100%, then the volatilization loss will go up to 31% during the rice growing season [70]. The NH_{4^+} concentration in surface water and the fertilization timing influences the NH_3 loss. A significant amount of loss in the first 1–7 days of fertilization then gradually the rate flattened [23]. The urea and NH_{4^+} enter 1–2 cm depth of soil surface within a week; thus, the broadcast urea should match the crop demand for maximum N recovery, and crops with superficial root systems benefit from absorbing NH_{4^+} [71]. Ammonia volatilization loss occurs after ammonium bicarbonate and urea application to flooded rice during transplanting, with losses of 39% and 30% of applied N, respectively [72].

The use of N-fertilizer in paddy water encourages algal growth [73]. Algae photosynthesize, removing CO_2 from the water and reducing the formation of carbonic acid. While daylight hours are ideal for this process, the paddy field water pH can rise as high as 9.0. These pH levels are the ideal conditions for NH_{4^+} compounds to release NH_3 into the air, and soil pH has a strong relation with NH_3 loss (**Figure 7a**). The crop's recovery or nitrogen use efficiency varies between 30% and 40%, and this mostly depends on the plant root's capability to drag N from the downward-moving pool of N [4]. The initial distribution of urea in the soil, hydrolysis rate, and N absorption rate by rice roots influences cumulative NH_3 volatilization. Adding organic matter in rice fields may increase CO_2 and reduce soil pH, but this will not work to reduce volatilization loss as the produced CO_2 may interact with the diurnal floodwater pH; however, the daily average soil pH remains unaffected.

It is defined that high soil pH (pH < 9) or alkaline soil conditions favor the volatilization rate. In contrast, volatilization loss is minimum in acidic soils but also can occur, especially volatilization loss occurs in calcareous soils when ammonium-containing fertilizers are applied on the soil surface. Basic soils containing $Ca(OH)_2$ may react with $(NH_4)_2CO_3$ to NH_4OH, easily decomposing to NH_3 and H_2O.

$$Ca(OH)_2 + (NH_4{}^+)_2CO_3 \longrightarrow NH_4OH + CaCO_3\ (ppt) \tag{6}$$

$$NH_4OH \longleftrightarrow NH_{3(g)} + H_2O \tag{7}$$

From the equations, it can be concluded that (1) at higher pH values, NH_3 volatilization is more pronounced, or (2) amending the solution with NH_3 gas-producing amendments will cause the reaction to move leftward, resulting in a rise in soil pH. Acidification is not caused by removing fertilizer N from the soil by crops as NH_{4^+}. After fertilization, low floodwater pH values in acidic soils resulted in (8−18%) volatilization loss, which is less than the denitrification loss (40−50% of applied N), where the total nitrogen loss ranges from 48% to 60% of applied N. Low solar radiation and poor algal growth at a rice paddy field in acid soils biosphere also facilitate low volatilization loss, in comparison with floodplain or calcareous soils [72]. It can be noted that only when the water entering the rice field differs from the concentration of acid or base from the water leaving the field does a permanent change in soil pH [4].

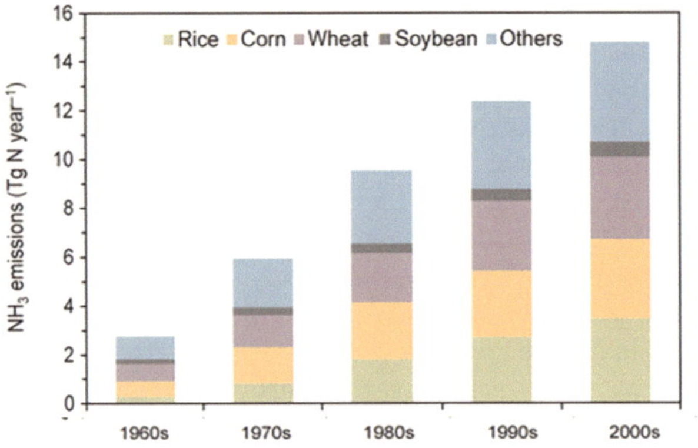

Figure 8.
Crop-specific NH_3 emissions from synthetic N fertilizer application [3].

Application of 71.4 Tg N year^{-1} as chemical fertilizer raised NH_3 emission from 2.8 ± 1.5 to 12.0 ± 0.8 Tg N year^{-1} [74]. The highest global mean NH_3 emission was estimated at 16.7 ± 0.5 Tg N year^{-1} in 2010. They also identified four major crops for NH_3 emission determination due to N fertilization, where in 2000 the largest NH_3 emitter was rice field (23.5%), followed by wheat (22.8%) then corn (21.9%), and the lowest emission was from soybean (<10%) field (**Figure 8**). The high emission from rice fields was due to increased N fertilizer application with increased cultivable area.

6. Gaseous losses of nitrogen under different water management techniques in rice field

6.1 Alternate wetting and drying (AWD)

Due to the continuous ponding of water in the rice field, the N transformation and transport processes are unique in rice ecosystem [75]. In rice field about 3600 l water Kg^{-1} grain is kept where 50–80% of the water loses through deep percolation [68], causing water and nitrogen loss from the field [76]. In conventional paddy cultivation, water requirement is high though the water is not fully used for the crop rather it facilitates nutrient loss. So, to save water resources, several water management techniques for rice fields are invented but among them, widely accepted technique is the alternate wetting and drying (AWD) irrigation practice [77]. Rice production is expected to shift from continuous flooding irrigation to AWD irrigation practice. Alternate wetting and drying is water-saving technique where the soil is not constantly flooded instead after the ponded water has receded, it is left to drain for one or more days before being re-flooded. The aerobic (dry) and anaerobic (wet) alternate conditions in the field exert complexity in N transformation processes. This increased yield and biomass over continuous flooding practice (**Figure 9**). In the 1980s and 1990s, research into AWD as a water-saving strategy began in China and India. The practice of AWD was originally tested as a water-saving strategy in the Philippines in 2002, and then in Bangladesh at the Bangladesh Rice Research Institute (BRRI) in 2005 [7]. But it is

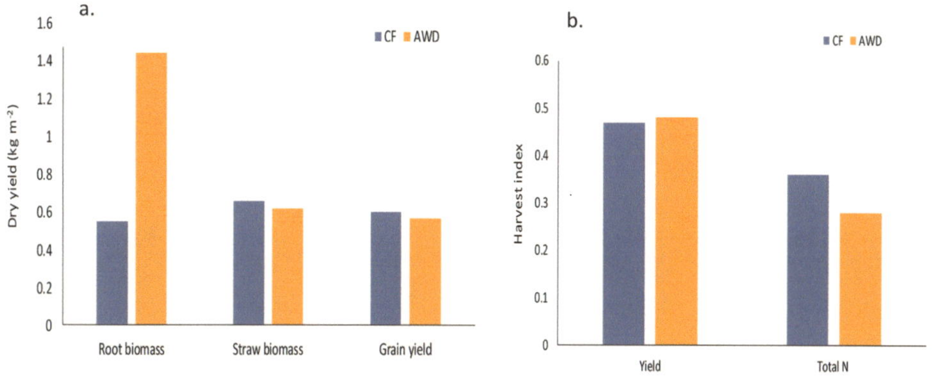

Figure 9.
Comparison of alternate wetting and drying (AWD) and continuous flooding (CF) practices on (a) root-straw biomass and grain yield (b) harvest index of yield and total N [78].

evident that this practice reduces CH_4 emission in the field by 38% [79] but there are also subsequent evidences that AWD increases N_2O emission [6, 80]. Conventional tillage practice reduces global warming potential (GWP) by 10–16% in comparison with new modern practices [81]. However, the decrease in CH_4 emissions brought about by AWD substantially balances the increase in N_2O emissions created.

Transformational modifications (anaerobic rice systems into aerobic) in rice cultivation practices sustain yield (**Figure 6a**) but at the cost of higher N loss [17] mostly N_2O. In comparison with conventional rice cultivation practice with water stress condition of −15 to −30 k Pa at 15–20 cm below the soil surface the grain yield is reduced by 11–32% [82, 83]. In Arkansas, it was stated that their less aggressive AWD treatment resulted in a 48% drop in methane emissions and that overall greenhouse gas (GHG) emissions were reduced by 45% when the increase in nitrous oxide was taken into consideration [79]. In AWD, the N_2O is prompted by enhanced denitrification of NO_{3-} during the rewetting of dry soils and the following increased nitrification of NH_4 during the dry phase [6]. During the drying phase of AWD irrigation practice, the soil aeration is good and this may improve the land quality, grain yield, utility of N, and water productivity along with reduction in N leaching, while N loss through ammonia volatilization and denitrification, and nitrification increased in AWD practice due to aerobic condition [84]. In comparison with conventional tillage, ammonia volatilization increases by 14% in medium moisture stress conditions and 17% in severe moisture stress conditions whereas denitrification increases by 7% in both medium and severe moisture stress conditions [61, 68]. **Table 5** shows a typical comparison:

By managing water table levels through controlled drainage and controlled irrigation.

Their effect on nitrification and volatilization can be determined. As the water table control levels increased, irrigation water volumes in the controlled irrigation paddy fields decreased. Seasonal ammonia volatilization losses reduce with the successive increase in controlled water table and the range varies from 53 kg N ha^{-1} in near to surface to 59 kg N ha^{-1} in below surface level [61]. The application of controlled drainage by raising water table to a suitable level could effectively reduce irrigation water volumes and ammonia volatilization losses from

Treatments	Annual emission of N_2O (kg N_2O-N ha^{-1})
Water seeded with conventional continuous flood irrigation	0.102
Water seeded with alternate wetting and drying (AWD)	0.142
Drill seeded with alternate wetting and drying (AWD)	0.616

Table 5.
Comparison between conventional irrigation and AWD based on N_2O emission in water seeded and drill seeded planting system [85].

paddy fields [61]. The combination of controlled irrigation and controlled drainage is a feasible water management method of reducing ammonia volatilization losses from paddy fields where NH_{4+} undergoes an oxidation process in dry spell that converts it to NO_{2-} and finally to nitrate NO_{3-}, where it is reduced to N_2 and nitrous oxides, which are also released into the atmosphere as a product of denitrification [86]. In AWD irrigation practice, the alternate aerobic and anaerobic conditions accelerate denitrification and volatilization process of N loss, whereas nitrification process is more favorable in anaerobic conditions. The redox potential under ponding condition is lower and this facilitates the N_2O formation *via* denitrification [81]. The NH_4^+ ions are higher in topsoil layers than in lower layers (10–70 cm) of soils throughout the rice growing season and this may occur due to (1) the presence of 3–5 cm thick plow pans below 15–20 cm of the soil surface which restricts the NH_{4+} ion movement from the top layers to the lower layers (**Figure 4**; [87]) and (2) the negatively charged soil colloid have high affinity to make bond with positively charged ions and restrict the ions from being lost. The AWD irrigation practice increases soil temperature. Temperature and soil moisture are the major factors for N transformation pathways where the high temperature is potential influencer for higher NO_{3-} concentration than NH_4^+ concentration due to denitrification and nitrification processes under soil moisture stress conditions in AWD [88]. In AWD irrigation practice, the ammonia volatilization and nitrification processes are increased compared to conventional rice cultivation practice [65]. Denitrification can turn ammonium into nitrite, nitric oxide, or nitrous oxide, which can contaminate groundwater or the atmosphere if certain conditions are met during nitrification. Furthermore, ammonium can react with nitrite in the soil to produce dinitrogen gas as a result of nitrite accumulation.

The N mineralization can go up to 75.5–80 kg ha^{-1} under no moisture stress condition where at 20–35% of the maximum water holding capacity of soil N mineralized by 55 to 64 kg ha^{-1} from 135 kg N ha^{-1} [65]. In dry conditions, OM decomposition leads to high ammonification of N, which is followed by NH3 loss during flooding. Aridity causes ammonification followed by nitrification-denitrification, resulting in the production of nitrogen dioxide and nitrous oxides again. In AWD irrigation practice, a large portion of N loss will occur in volatilization process accounting to 21% of the applied N and continuous flooding cause 13% N loss of the applied N. The NH_3/NH_{4+} ratio was largely determined by soil and floodwater pH, which influenced ammonia volatilization. Higher irrigation water levels can reduce ammonia losses, because of an NH_{4+} dilution effect [89]. Nitrification-denitrification losses of fertilizer-N are six times greater under AWD than continuous flooding [78].

6.2 Dry direct seeded rice (DSR) planting system

In contrast to transplanting seedlings from the nursery, direct seeding of rice refers to the practice of starting the rice crop from seeds which are put directly in the field [18]. The three main direct seeding techniques for rice (DSR) are water seeding (seeds sown into standing water), wet seeding (sowing pregerminated seeds on wet puddle soils), and dry seeding (Sowing dry seeds into dry soil [18]). Dry seeding has been practiced as a principal method of rice establishment in developed countries since 1950s [90]. Puddling limits soil permeability and produces a hard pan below the plow-zone in the classic transplanting system (TPR). Due to percolation, surface evaporation, and puddling, there are significant water losses as a result. Contrary to puddle transplanting, dry seeding enables the production of dry season (Boro) rice with less than 50% of the irrigation water needed [18]. But there is a significant matter to consider here. Since the field is supposed to be dry for a long period of time while practicing DSR technique, there is a chance that it would emit N_2O (**Table 6**). It is quite an established fact that dry farming increases N_2O production. It is found that DSR technique causes N_2O emissions, and it cannot be neglected.

7. Strategies to reduce GHGs emission from rice fields

Agriculture is the major source and worse victim of climate change effects. Paddy field emits CH_4 and N_2O which are major GHGs due to their high GWP and it was estimated that agriculture contributes 52% CH_4 and 84% N_2O anthropogenic GHG emission. The agriculture is also subject to 20–40% soil organic matter loss due to cultivation [15]. As consequence, the nutrient holding capacity of soil is also reduced. The wise use of agricultural technologies and practices are potential source to reduce GHG emission from cropland and other agricultural sectors. If the modern strategies are successfully applied, they are more likely to boost crop and animal production than to diminish it (**Table 7**). A list of probable practices has given below:

District	Emission of N_2O (t CO_2 eq. ha^{-1}) in conventional puddled transplanted rice	Emission of N_2O (t CO_2 eq. ha^{-1}) in dry direct seeded rice planting system
Amritsar	0.4	0.6
Barnala	0.4	0.6
Bathinda	0.3	0.5
Faridcot	0.5	0.6
Fatehgarh Sahib	0.5	0.6
Ferozepur	0.5	0.6
Gurdaspur	0.4	0.5

Table 6.
Emission of nitrous oxide (N_2O) under conventional and direct-seeded rice in different districts of Punjab, India [91].

Country	Location	Water regime[a]	Nitrogen fertilizer[b]	Tested mitigation option[c]	Amount of chemical N, kg N ha^{-1}	Emission of tested mitigation option plot,[d] %	Reference
India	New Delhi	MSD	U	DCD	140	84[e]	Kumar et al. [56]
India	New Delhi	MSD	AC	DCD	140	68[e]	Kumar et al. [56]
India	New Delhi	MSD	U	NUE	140	89	Majumdar et al. [57]
India	New Delhi	CF	PN	DCD	120	87[e]	Ghosh et al. [47]
India	New Delhi	CF	U	DCD	120	39[e]	Pathak et al. [48]
Philippines	Los Banos	RF	U	Slow-U	90	3[e,f]	Abao et al. [50]

[a]*Water regime: MSD, midseason drainage; CF, continuous flooding; RF, rain-fed, wet-season.*
[b]*Fertilizer type: AS, ammonium sulfate; PN, potassium nitrate; U, urea.*
[c]*Mitigation options: DCD, dicyandiamide; NEU, neem-coated urea; TS, thiosulfate; slow-U, slow-release urea.*
[d]*Fertilizer-induced N_2O-N emission of the tested mitigation option plot compared with that of the conventional fertilizer plot.*
[e]*Significantly different from conventional fertilizer plot at $P < 0.05$ by Duncan's multiple range test. Statistical test results are from the original papers.*
[f]*Fertilizer-induced N_2O emission could not be calculated because no zero-N control plot was available. Thus, the percent of N_2O-N emission (including background emission) from the tested mitigation option plot is compared with that from conventional fertilizer plot is shown.*

Table 7.
Available data on possible mitigation options [44].

7.1 Ways of reducing ammonia volatilization

7.1.1 Salts of potassium

When urea is added to soil, it converts to ammonium carbonate, which is prone to NH_3 volatilization loss. When calcium or magnesium nitrate or chloride salts are combined with urea, they minimize volatilization by generating ammonium chloride or nitrate [92]. Potassium (K) indirectly reduces NH_3 volatilization loss by enhancing calcium carbonate precipitation in high Ca soil by replacing Ca from the exchange complex. Potassium nitrate (KNO_3) or potassium chloride (KCl) might be utilized to reduce NH_3 volatilization losses [93]. When Mg, Ca, or K were combined with urea, the ammonia loss from urea decreased with increasing cation/N ratio [94].

7.1.2 Inhibitors

When urea is added to the soil, the urease enzyme hydrolyzes it and converts it to ammonium carbonate. The use of urease inhibitors reduces urease activity at the soil surface, allowing urea to migrate deeper into the soil before hydrolysis. The urease inhibitors phenyl phosphor diamidate (PPD) and N-(n-butyl) thio phosphoric triamide (NBPT) were successful in lowering ammonia volatilization loss in laboratory and greenhouse trials [95, 96]. Freney et al. [97] discovered that using an algal inhibitor (copper sulfate terbutryn) reduced ammonia volatilization loss, resulting in a 0.3–0.6 t ha improvement in rice output. Rawluk et al. [98] reported that the use of NBPT with granular urea reduced ammonia volatilization loss by 28–88% [93].

7.2 Denitrification minimization

Through the nitrification process, the ammonium ion remaining in the soil-water system is easily transformed to nitrite, then to nitrate. The nitrate ion is lost due to denitrification and leaching. The nitrification process is followed by denitrification. Denitrification loss will be decreased if ammonium nitrification into nitrate is delayed or reduced. As a result, nitrification inhibitors, such as dicyandiamide (DCD), iron pyrite, nitrapyrin, phenylacetylene, encapsulated calcium carbide, terrazole, and others, can be used to reduce denitrification losses [93, 99]. One of the key three greenhouse gases is nitrous oxide, which is emitted from agricultural soils owing to denitrification loss (methane, carbon dioxide, and nitrous oxide). The use of plant residues with high polyphenol content and protein binding capacity may help to minimize nitrous oxide emissions [93, 100].

7.3 Biochar addition

When tested on 14 different agricultural soils, biochar was shown to reduce denitrification and N_2O emissions by 10–90%, with a consistent reduction of the $N_2O/(N_2 + N_2O)$ ratio, indicating that biochar reduces N_2O emissions by facilitating the last step of the denitrification process and producing more N_2 rather than N_2O [101]. However, biochar application in some soils can accelerate nitrification and increase N_2O emissions; hence, the effect of biochar application on N^2O emissions is connected to the primary N_2O production pathway that runs in a soil [102]. Uzoma et al. [103] said the increase in anion exchange capacity of biochar reduces leaching of anionic (NO_3^-) nutrients while the cation exchange capacity increases the adsorption of cation (NH_4^+) nutrients. Therefore, this implies that application of inorganic fertilizer N alongside biochar improves retention and uptake of both NO_3^- and NH_4^+.

7.4 Integrated nutrient management

Enhancing NUE can be accomplished by integrated nutrient management, which includes the use of organic manures, green manures, legumes, agricultural wastes, and biofertilizers. Organic manures are an additional source of nutrients and improve the effectiveness of fertilizers. Combining the use of organic manure and nitrogen fertilizer helps provide a steady supply of nitrogen, reduces loss, and improves the application of nitrogen [104]. Rice cv. Pusa Basmati-1 recovered more nitrogen from the fertilizer when N was applied, half as urea and half as FYM on a sandy loam soil [105]. The partial factor productivity of nitrogen (PFPN) values for the rice-rice system ranged from 26 to 52 kg grain kg^{-1} N under recommended NPK, but they increased to 33–77 kg grain kg^{-1} N with the substitution of 25% N through FYM in kharif rice and the reduction of 25% N in succeeding rabi rice (**Table 8**). Due to the fixation of atmospheric N, green manuring with legumes enhances soil N. GM enhances the physical, chemical, and biological characteristics of soil in addition to

Crop	N rate (kg ha^{-1})		Grain yield (tn ha^{-1})		AE_N	
	FNP	INM	FNP	INM	FNP	INM
Rice	167	135	5.90	6.50	9	16
Wheat	325	130	5.76	6.02	3	11
Maize	263	158	8.45	8.90	5	11

Table 8.
Grain yield and nitrogen recovery of different cereals as influenced by integrated nitrogen management [106].

minimizing leaching and gaseous losses of N. Sunnhemp (*Crotolaria juncea*) and dhaincha (*Sesbania aculeata*) are the most common GM crops. Incorporating legumes could provide an average of 50–60 kg N ha^{-1}, according to an evaluation of a variety of leguminous crops for satisfying the N demand of a succeeding nonlegume crop [104].

7.5 Slow-release fertilizer

Nitrate-containing fertilizers are subjected to leaching while ammonium and amide-containing nitrogenous fertilizer are more susceptible to volatilization loss. Slow-release fertilizers can reduce the nitrogen loss by delaying nitrogen release and enhancing the synchronization between crop demand and soil nitrogen supply [107]. These compounds are produced by treating highly soluble urea fertilizer with substances that prevent or slow down the fertilizer's hydrolysis to ammonium.

For example, urea–formaldehyde, isobutylidene diurea (IBDU), resin-coated fertilizers (e.g., Osmocote®), polymer and sulfur-coated urea. (brady) The nitrogen uptake for the treatments of BU (Bare Urea), CRU (Controlled release urea), MBC (50% BU + 50% CRU) and MBCB (50% BU + 50% CRU + biochar) significantly increased by 28.3%, 73.0%, 80.0% and 91.1% over that of the CK, respectively [108]. Nitrogen recovery and yield of basmati rice under different slow-release nitrogenous fertilizers are in **Table 9**.

Slow-release urea enhanced single rice yield by 6.0–31.2% and NUE by 20.3–96.5% compared to the same amount of conventional urea or slow-release urea applied alone when slow-release N comprised 30–70% of the total N [110].

The maximum possibility of N fertilizer contributing to the environmental effect (MPEI) would be lowered by 67% when NUE were to increase by 50% and the N fertilization rate were to decrease by 34% [111].

N source	N rate (kg ha^{-1})	Grain yield (q ha^{-1})	N uptake (kg ha^{-1})	AE_N
PU	120	32.9	70.2	3.47
PNGU	120	36.2	82.7	6.25
KEU	120	32.5	71.9	3.13
N Sources: PU—Prilled urea, PNGU—Pusa Neem Golden Urea, KEU—Karanj Coated Urea.				

Table 9.
Nitrogen recovery and yield of basmati rice under different slow-release nitrogenous fertilizers [109].

7.6 Nitrification inhibitors

Inhibitors of nitrification are substances that prevent the Nitrosomonas bacteria from converting NH_{4+} to NO_{2-} in the first phase of nitrification and that will raise NUE and crop output by ensuring a larger concentration of ammoniacal nitrogen in the soil medium [109]. Nitrification inhibitors including dicyandiamide (DCD), nitrapyrin (N-Serve®), 3, 4-Dimethylpyrazole phosphate (DMPP), Ca-carbide, and etridiazol (Dwell®) have been developed by chemical companies. These substances can temporarily stop the generation of NO_3 when combined with nitrogen fertilizers. Temporarily is the essential word here, as the inhibition typically only lasts a few weeks (less if soil temperatures are over 20°C) when nitrification conditions are good [73].

8. Conclusion

Adaptation of modern agricultural practices, such as DSR and AWD, somehow reduce NH_3 and CH_4 emission but increase N_2O emissions from rice fields. To off-set gaseous losses of N from rice field with environmental benefits IPNS-based strategies are highly required and deep placement of urea is also a possible method. Soil is a heterogenous body with complex ecosystem where the interactions among soil properties (pH, SOC, TN, and mineral N) are associated with the mitigation of the GWP from N_2O and CH_4 emissions for sustainable agriculture.

Author details

Jannatul Ferdous, Farah Mahjabin, Mohammad Abdullah al Asif, Israt Jahan Riza and Mohammad Mofizur Rahman Jahangir*
Bangladesh Agricultural University, Mymensingh, Bangladesh

*Address all correspondence to: mmrjahangir@bau.edu.bd

References

[1] Haque AN, Uddin MK, Sulaiman MF, Amin AM, Hossain M, Solaiman ZM, et al. Biochar with alternate wetting and drying irrigation: A potential technique for paddy soil management. Agriculture. 2021;**11**(4):367

[2] Uddin S, Nitu TT, Milu UM, Nasreen SS, Hossenuzzaman M, Haque ME, et al. Ammonia fluxes and emission factors under an intensively managed wetland rice ecosystem. Environmental Science: Processes & Impacts. 2021;**23**(1):132-143

[3] Xu R, Tian H, Pan S, Prior SA, Feng Y, Batchelor WD, et al. Global ammonia emissions from synthetic nitrogen fertilizer applications in agricultural systems: Empirical and process-based estimates and uncertainty. Global Change Biology. Jan 2019;**25**(1):314-326

[4] Kirk G. The Biogeochemistry of Submerged Soils. John Wiley & Sons; 2004

[5] IFA (International Fertilizer Association). Fertilizer Outlook 2019–2023. 87th IFA Annual Conference Summary Report, Montreal, Canada, 11–13 June 2019. Paris; 2019. Available from: http://www.fertilizer.org

[6] Islam SM, Gaihre YK, Islam MR, Akter M, Al Mahmud A, Singh U, et al. Effects of water management on greenhouse gas emissions from farmers' rice fields in Bangladesh. Science of the Total Environment. 2020;**734**:139382

[7] Lampayan RM, Samoy-Pascual KC, Sibayan EB, Ella VB, Jayag OP, Cabangon RJ, et al. Effects of alternate wetting and drying (AWD) threshold level and plant seedling age on crop performance, water input, and water productivity of transplanted rice in Central Luzon, Philippines. Paddy and Water Environment. 2015;**13**(3):215-227

[8] Nitu TT, Milu UM, Jahangir MM. Cover crops and soil nitrogen cycling. In Cover Crops and Sustainable Agriculture. CRC Press; 2021. pp. 209-226

[9] Fillery IR, Vlek PL. The significance of denitrification of applied nitrogen in fallow and cropped rice soils under different flooding regimes I. Greenhouse experiments. Plant and Soil. Jun 1982;**65** (2):153-169

[10] Singh B, Singh Y. Reactive nitrogen in Indian agriculture: Inputs, use efficiency and leakages. Current Science. 2008;**94**(11):00113891

[11] Zheng X et al. The Asian nitrogen cycle case study. Ambio. 2002;**31**:79-87

[12] Ma J, Ma E, Xu H, Yagi K, Cai Z. Wheat straw management affects CH_4 and N_2O emissions from rice fields. Soil Biology and Biochemistry. 2009;**41**(5): 1022-1028

[13] IPCC. Agriculture, forestry and other land use (AFOLU). In: Edenhofer O, Pichs-Madruga R, Sokona Y, Farahani E, Kadner S, Seyboth K, et al., editors. Climate Change 2014: Mitigation of Climate Change. Contribution of Working Group III to the Fifth Assessment Report of the Intergovernmental Panel on Climate Change. Cambridge: Cambridge University Press; 2015. pp. 811-922. DOI: 10.1017/cbo9781107415416.017

[14] Hoffman FM, Randerson JT, Arora VK, Bao Q, Cadule P, Ji D, et al. Causes and implications of persistent atmospheric carbon dioxide biases in earth system models. Journal of Geophysical Research—Biogeosciences. 2014;**119**:141-162. DOI: 10.1002/2013JG002381

[15] Zaman M, Kleineidam K, Bakken L, Berendt J, Bracken C, Butterbach-Bahl K, et al. Measuring Emission of Agricultural Greenhouse Gases and Developing Mitigation Options Using Nuclear and Related Techniques. Springer Nature; 2021. DOI: 10.1007/978-3-030-55396-8. ISBN 978–3–030-55395-1

[16] Hwalla N, El Labban S, Bahn RA. Nutrition security is an integral component of food security. Frontiers in Life Science. 2016;**9**:167-172. DOI: 10.1080/21553769. 2016.1209133

[17] Farooq MS, Uzair M, Maqbool Z, Fiaz S, Yousuf M, Yang SH, et al. Improving nitrogen use efficiency in aerobic rice based on insights into the ecophysiology of archaeal and bacterial ammonia oxidizers. Frontiers in Plant Science. 2022;**13**

[18] Rahman MM, Uddin M. Organic amendment on soil quality and yield performance of dry direct seeded Boro Rice. Bangladesh Agronomy Journal. 2020a;**23**(2):103-109

[19] Mohanty S, Nayak AK, Swain CK, Dhal BR, Kumar A, Kumar U, et al. Impact of integrated nutrient management options on GHG emission, N loss and N use efficiency of low land rice. Soil and Tillage Research. 2020;**200**:104616

[20] Fillery IR, Simpson JR, De Datta SK. Influence of field environment and fertilizer management on ammonia loss from flooded rice. Soil Science Society of America Journal. 1984;**48**(4):914-920

[21] Simpson JR, Freney JR, Wetselaar R, Muirhead WA, Leuning R, Denmead OT. Transformations and losses of urea nitrogen after application to flooded rice. Australian Journal of Agricultural Research. 1984;**35**:189-200

[22] Chen JL, Ortiz R, Steele TW, Stuckey DC. Toxicants inhibiting anaerobic digestion: A review. Biotechnology Advances. 2014;**32**(8):1523-1534

[23] Rahaman MA, Zhan X, Zhang Q, Li S, Long Y, Zeng H. Ammonia volatilization reduced by combined application of biogas slurry and chemical fertilizer in maize–wheat rotation system in North China plain. Sustainability. 2020;**12**(11):4400

[24] Liu SK, Cai S, Chen Y, Xiao B, Chen P, Xiang XD. The effect of pollutional haze on pulmonary function. Journal of Thoracic Disease. 2016;**8**(1):E41

[25] Prasad R, Shivay YS, Kumar D. Current status, challenges, and opportunities in rice production. Rice Production Worldwide. 2017:1-32

[26] Hayatsu M, Tago K, Saito M. Various players in the nitrogen cycle: Diversity and functions of the microorganisms involved in nitrification and denitrification. Soil Science and Plant Nutrition. 2008;**54**(1):33-45

[27] Guggenberger G. Humification and mineralization in soils. In Microorganisms in Soils: Roles in Genesis and Functions. Berlin, Heidelberg: Springer; 2005. pp. 85-106

[28] Kemp MJ, Dodds WK. The influence of ammonium, nitrate, and dissolved oxygen concentrations on uptake, nitrification, and denitrification rates associated with prairie stream substrata. Limnology and Oceanography. 2002;**47**(5):1380-1393

[29] Nieder R, Benbi DK. Carbon and nitrogen transformations in soils. Carbon and Nitrogen in the Terrestrial

Environment. 2008:137-159. DOI: 10.1007/978-1-4020-8433-1

[30] Abrol YP, Raghuram N, Sachdev MS, editors. Agricultural Nitrogen Use and its Environmental Implications. New Delhi: IK International Pvt Ltd; 2007

[31] Hu BL, Shen LD, Xu XY, Zheng P. Anaerobic ammonium oxidation (anammox) in different natural ecosystems. Biochemical Society Transactions. 2011;**39**(6):1811-1816

[32] Zhou H, Zhang Q, Gu C, Jabeen S, Li J, Di H. Soil anammox community structure in different land use soils treatment with 13C urea as determined by analysis of phospholipid fatty acids. Applied Microbiology and Biotechnology. 2017;**101**(17):6659-6669

[33] De Datta SK, Samson MI, Obcemea WN, Real JG, Buresh RJ. Direct measurement of ammonia and denitrification fluxes from urea applied to rice. Soil Science Society of America Journal. 1991;**55**:543e548

[34] Zhao X, Xie YX, Xiong ZQ, Yan XY, Xing GX, Zhu ZL. Nitrogen fate and environmental consequence in paddy soil under rice-wheat rotation in the Taihu lake region, China. Plant and soil. 2009;**319**(1):225-234

[35] Pathak H, Prasad S, Bhatia A, Singh S, Kumar S, Singh J, et al. Methane emission from rice–wheat cropping system in the Indo-Gangetic plain in relation to irrigation, farmyard manure and dicyandiamide application. Agriculture, Ecosystems & Environment. 2003;**97**(1–3):309-316

[36] Song YS, Fan XH, Lin DX, Yang LZ, Zhou JM. Investigation on ammonia loss from the flooded rice field in Taihu region and its influencing factors. Acta Pedologica Sinica. 2004;**41**(2):265-269

[37] Liang XQ, Chen YX, Li H, Tian GM, Ni WZ, He MM, et al. Modeling transport and fate of nitrogen from urea applied to a near-trench paddy field. Environmental Pollution. 2007;**150**(3): 313-320

[38] Fowler D, Steadman CE, Stevenson D, Coyle M, Rees RM, Skiba UM, et al. Effects of global change during the 21st century on the nitrogen cycle. Atmospheric Chemistry and Physics. 2015;**15**(24):13849-13893

[39] Tesfai M. Emissions of N2O from agricultural soils and mitigation options: A review with special reference to Norwegian agriculture. 2016;**2**:25

[40] Chen WY, Suzuki T, Lackner M, editors. Handbook of Climate Change Mitigation and Adaptation. Gewerbestrasse, Cham, Switzerland: Springer International Publishing; 2017

[41] Buresh RJ, Ramesh Reddy K, Van Kessel C. Nitrogen transformations in submerged soils. Nitrogen in Agricultural Systems. 2008;**49**:401-436

[42] Ponnamperuma FN. Effects of Flooding on Soils. New York: Academic Press; 1984

[43] Zou J, Huang Y, Jiang J, Zheng X, Sass RL. A 3-year field measurement of methane and nitrous oxide emissions from rice paddies in China: Effects of water regime, crop residue, and fertilizer application. Global Biogeochemical Cycles. 2005;**19**(2)

[44] Akiyama H, Yagi K, Yan X. Direct N_2O emissions from rice paddy fields: Summary of available data. Global Biogeochemical Cycles. 2005;**19**(1)

[45] Xu YC, Shen QR, Li ML, Dittert K, Sattelmacher B. Effect of soil water status and mulching on N_2O and CH4

emission from lowland rice field in China. Biology and Fertility of Soils. 2004;**39**(3):215-217

[46] Chen GX, Huang GH, Huang B, Yu KW, Wu J, Xu H. Nitrous oxide and methane emissions from soil–plant systems. Nutrient Cycling in Agroecosystems. 1997;**49**(1):41-45

[47] Ghosh S, Majumdar D, Jain MC. Methane and nitrous oxide emissions from an irrigated rice of North India. Chemosphere. 2003;**51**(3):181-195

[48] Pathak H, Bhatia A, Prasad S, Singh S, Kumar S, Jain MC, et al. Emission of nitrous oxide from rice-wheat systems of Indo-Gangetic plains of India. Environmental Monitoring and Assessment. 2002 Jul;**77**(2):163-178

[49] Suratno W, Murdiyarso D, Suratmo FG, Anas I, Saeni MS, Rambe A. Nitrous oxide flux from irrigated rice fields in West Java. Environmental Pollution. 1998;**102**(1):159-166

[50] Abao EB, Bronson KF, Wassmann R, Singh U. Simultaneous records of methane and nitrous oxide emissions in rice-based cropping systems under rainfed conditions. In Methane Emissions from Major Rice Ecosystems in Asia 2000 (pp. 131-139). Springer, Dordrecht.

[51] Smith CJ, Brandon M, Patrick WH Jr. Nitrous oxide emission following urea-N fertilization of wetland rice. Soil Science and Plant Nutrition. 1982;**28**(2):161-171

[52] Cai ZC, Xing GX, Shen GY, Xu H, Yan XY, Tsuruta H, et al. Measurements of CH_4 and N_2O emissions from rice paddies in Fengqiu, China. Soil Science and Plant Nutrition. 1 Mar 1999;**45**(1):1-3

[53] Xiong ZQ, Xing GX, Tsuruta H, Shen GY, Shi SL, Du LJ. Measurement of nitrous oxide emissions from two rice-based cropping systems in China. Nutrient Cycling in Agroecosystems. 2002;**64**(1):125-133

[54] Khalil MA, Rasmussen RA, Shearer MJ, Chen ZL, Yao H, Yang J. Emissions of methane, nitrous oxide, and other trace gases from rice fields in China. Journal of Geophysical Research: Atmospheres. 1998;**103**(D19):25241-25250

[55] Zheng X, Wang M, Wang Y, Shen R, Gou J, Li J, et al. Impacts of soil moisture on nitrous oxide emission from croplands: A case study on the rice-based agro-ecosystem in Southeast China. Chemosphere-Global Change Science. 2000;**2**(2):207-224

[56] Kumar U, Jain MC, Pathak H, Kumar S, Majumdar D. Nitrous oxide emission from different fertilizers and its mitigation by nitrification inhibitors in irrigated rice. Biology and Fertility of Soils. 2000;**32**(6):474-478

[57] Majumdar D, Kumar S, Pathak H, Jain MC, Kumar U. Reducing nitrous oxide emission from an irrigated rice field of North India with nitrification inhibitors. Agriculture, Ecosystems & Environment. 2000;**81**(3):163-169

[58] Hou AX, Chen GX, Wang ZP, Van Cleemput O, Patrick WH Jr. Methane and nitrous oxide emissions from a rice field in relation to soil redox and microbiological processes. Soil Science Society of America Journal. 2000;**64**(6):2180-2186

[59] Nishimura S, Sawamoto T, Akiyama H, Sudo S, Yagi K. Methane and nitrous oxide emissions from a paddy field with Japanese conventional water management and fertilizer application. Global Biogeochemical Cycles. 2004;**18**(2)

[60] Tsuruta H, Kanda K, Hirose T. Nitrous oxide emission from a rice paddy field in Japan. Nutrient Cycling in Agroecosystems. 1997;**49**(1):51-58

[61] He Y, Yang S, Xu J, Wang Y, Peng S. Ammonia volatilization losses from paddy fields under controlled irrigation with different drainage treatments. The Scientific World Journal. 2014;**2014**. Article ID 417605. DOI: 10.1155/2014/417605

[62] IPCC. Chapter 11 N_2O emissions from managed soils, and CO_2 emissions from lime and urea application. In: 2006 IPCC Guidelines for National Greenhouse Gas Inventories. 2006

[63] Riddick S, Ward D, Hess P, Mahowald N, Massad R, Holland E. Estimate of changes in agricultural terrestrial nitrogen pathways and ammonia emissions from 1850 to present in the Community Earth System Model. Biogeosciences. 13 Jun 2016;**13**(11):3397-3426

[64] Bouwman AF, Boumans LJ, Batjes NH. Estimation of global NH_3 volatilization loss from synthetic fertilizers and animal manure applied to arable lands and grasslands. Global Biogeochemical Cycles. Jun 2002;**16**(2):8-1

[65] Tan X, Shao D, Gu W, Liu H. Field analysis of water and nitrogen fate in lowland paddy fields under different water managements using HYDRUS-1D. Agricultural Water Management. 2015;**150**:67-80

[66] Darzi-Naftchali A, Shahnazari A, Karandish F. Nitrogen loss and its health risk in paddy fields under different drainage managements. Paddy and Water Environment. 2017;**15**(1):145-157

[67] Li Y, Šimůnek J, Zhang Z, Jing L, Ni L. Evaluation of nitrogen balance in a direct-seeded-rice field experiment using Hydrus-1D. Agricultural Water Management. 2015;**148**:213-222

[68] Shekhar S, Mailapalli DR, Raghuwanshi NS. Simulating nitrogen transport in paddy crop irrigated with alternate wetting and drying practice. Paddy and Water Environment. 2021;**19**(3):499-513

[69] Dash CJ, Sarangi A, Singh DK, Singh AK, Adhikary PP. Prediction of root zone water and nitrogen balance in an irrigated rice field using a simulation model. Paddy and Water Environment. 2015;**13**(3):281-290

[70] Peng S, Hou H, Xu J, Mao Z, Abudu S, Luo Y. Nitrous oxide emissions from paddy fields under different water managements in Southeast China. Paddy and Water Environment. 2011;**9**(4):403-411

[71] Peng S, Cassman KG. Upper threshholds of nitrogen uptake rates and associated nitrogen fertilizer efficiencies in irrigated rice. Agronomy Journal. 1998;**90**(2):178-185

[72] Zhu ZL, Cai GX, Simpson JR, Zhang SL, Chen DL, Jackson AV, et al. Processes of nitrogen loss from fertilizers applied to flooded rice fields on a calcareous soil in north-Central China. Fertilizer Research. 1989;**18**:101e115

[73] Weil RR, Brady NC. The Nature and Properties of Soils. 15th ed. London, New York, Boston, San Francisco, Pearson, Harlow: England; 2017

[74] Xu X, Tian H, Hui D. Convergence in the relationship of CO_2 and N_2O exchanges between soil and atmosphere within terrestrial ecosystems. Global Change Biology. 2008;**14**(7):1651-1660

[75] Sutradhar B. Design and development of an institutional repository at the Indian Institute of Technology Kharagpur. Program Electronic Library and Information Systems. 2006;**40**(3):244-255

[76] Cai GX, Zhu ZL, Trevitt ACF, Freney JR, Simpson JR. Nitrogen loss from ammonium bicarbonate and urea fertilizers applied to flooded rice. Fertiliser Research. 1986;**10**:203-215

[77] Shekhar S, Mailapalli DR, Das BS, Raghuwanshi NS. Modelling Water Flow through Paddy Soils under Alternate Wetting and Drying Irrigation Practice. New Orleans, LA: AGUFM; 2017. pp. H43Q-H4H7

[78] Dong NM, Brandt KK, Sørensen J, Hung NN, Van Hach C, Tan PS, et al. Effects of alternating wetting and drying versus continuous flooding on fertilizer nitrogen fate in rice fields in the Mekong Delta, Vietnam. Soil Biology and Biochemistry. 2012;**47**:166-174

[79] Pittelkow CM, Liang X, Linquist BA, Van Groenigen KJ, Lee J, Lundy ME, et al. Productivity limits and potentials of the principles of conservation agriculture. Nature. 2015;**517**(7534):365-368

[80] Islam SM, Gaihre YK, Biswas JC, Jahan MS, Singh U, Adhikary SK, et al. Different nitrogen rates and methods of application for dry season rice cultivation with alternate wetting and drying irrigation: Fate of nitrogen and grain yield. Agricultural Water Management. 2018;**196**:144-153

[81] Jahangir MM, Bell RW, Uddin S, Ferdous J, Nasreen SS, Haque ME, et al. Conservation agriculture with optimum fertilizer nitrogen rate reduces GWP for rice cultivation in floodplain soils. Frontiers in Environmental Science. 2022:291

[82] Zhang H, Xue Y, Wang Z, Yang J, Zhang J. An alternate wetting and moderate soil drying regime improves root and shoot growth in rice. Crop Science. 2009;**49**(6):2246-2260

[83] Zhang H, Xue Y, Wang Z, Yang J, Zhang J. An alternate wetting and moderate soil drying regimes improves root and shoot growth in rice. Crop Science. 2009b;**49**:2246e2260

[84] Darzi-Naftchali A, Karandish F, Šimůnek J. Numerical modeling of soil water dynamics in subsurface drained paddies with midseason drainage or alternate wetting and drying management. Agricultural Water Management. 2018;**197**:67-78

[85] LaHue GT, Chaney RL, Adviento-Borbe MA, Linquist BA. Alternate wetting and drying in high yielding direct-seeded rice systems accomplishes multiple environmental and agronomic objectives. Agriculture, Ecosystems & Environment. 2016;**229**:30-39

[86] Patrick WH, Reddy KR. Nitrification-denitrification reactions in flooded soils and water bottoms: Dependence on oxygen supply and ammonium diffusion. Journal of Environmental Quality. 1976;**5**: 469-472

[87] Shekhar S, Mailapalli DR, Raghuwanshi NS, Das BS. Hydrus-1D model for simulating water flow through paddy soils under alternate wetting and drying irrigation practice. Paddy and Water Environment. Jan 2020;**18**(1): 73-85

[88] Tan X, Shao D, Gu W. Effects of temperature and soil moisture on gross nitrification and denitrification rates of a Chinese lowland paddy field soil. Paddy and Water Environment. 2018;**16**(4): 687-698

[89] Win KT, Nonaka R, Toyota K, Motobayashi T, Hosomi M. Effects of option mitigating ammonia volatilization on CH_4 and N_2O emissions from a paddy field fertilized with anaerobically digested cattle slurry. Biology and Fertility of Soils. 2010;**46**(6):589-595

[90] Zheng M, Tao Y, Hussain S, Jiang Q, Peng S, Huang J, et al. Seed priming in dry direct-seeded rice: Consequences for emergence, seedling growth and associated metabolic events under drought stress. Plant Growth Regulation. Mar 2016;**78**(2):167-178

[91] Pathak H, Sankhyan S, Dubey DS, Bhatia A, Jain N. Dry direct-seeding of rice for mitigating greenhouse gas emission: Field experimentation and simulation. Paddy and Water Environment. 2013;**11**(1):593-601

[92] Fenn LB, Taylor RM, Matocha JE. Ammonia losses from surface-applied nitrogen fertilizer as controlled by soluble calcium and magnesium: General theory. Soil Science Society of America Journal. 1981;**45**(4):777-781

[93] Choudhury AT, Kennedy IR. Nitrogen fertilizer losses from rice soils and control of environmental pollution problems. Communications in Soil Science and Plant Analysis. 2005;**36**(11–12):1625-1639. DOI: 10.1081/CSS-200059104. Copyright # Taylor & Francis, Inc. ISSN 0010-3624 print/1532-2416 online

[94] Khanif YM, Wong LP. Control of volatilization loss by soluble cations and urease inhibitors. In: Pushparajah E, Husin A, Bachik AT, editors. Proceedings of International Symposium on Urea Technology and Utilization. Taylor & Francis, Inc; 1987. pp. 311-320

[95] Byrnes BH, Savant NK, Craswell ET. Effect of a urease inhibitor phenyl phosphorodiamidate on the efficiency of urea applied to rice. Soil Science Society of America Journal. 1983;**47**(2):270-274

[96] Cai GX, Freney JR, Muirhead WA, Simpson JR, Chen DL, Trevitt AC. The evaluation of urease inhibitors to improve the efficiency of urea as a N-source for flooded rice. Soil Biology and Biochemistry. 1989;**21**(1):137-145

[97] Freney JR, Keerthisinghe DG, Phongpan S, Chaiwanakupt P, Harrington KJ. Effect of urease, nitrification and algal inhibitors on ammonia loss and grain yield of flooded rice in Thailand. Fertilizer Research. 1994;**40**(3):225-233

[98] Rawluk CD, Grant CA, Racz GJ. Ammonia volatilization from soils fertilized with urea and varying rates of urease inhibitor NBPT. Canadian Journal of Soil Science. 2001;**81**(2):239-246

[99] De Datta SK. Principles and Practices of Rice Production. A Wiely-Interscience Publication. International Rice Research Institute; 1981

[100] Baggs EM, Cadisch G, Verchot LV, Millar N, Ndufa JK. Environmental impacts of tropical agricultural systems: N_2O emissions and organic matter management. In: 17. World Congress of Soil Science, 14-21 August 2002. Bangkok (Thailand): FAO; 2002

[101] Cayuela ML, Sánchez-Monedero MA, Roig A, Hanley K, Enders A, Lehmann J. Biochar and denitrification in soils: When, how much and why does biochar reduce N_2O emissions? Scientific Reports. 2013;**3**(1):1-7

[102] Sánchez-García M, Roig A, Sánchez-Monedero MA, Cayuela ML. Biochar increases soil N2O emissions produced by nitrification-mediated

pathways. Frontiers in Environmental Science. 2014;**2**:25

[103] Uzoma KC, Inoue M, Andry H, Fujimaki H, Zahoor A, Nishihara E. Effect of cow manure biochar on maize productivity under sandy soil condition. Soil Use and Management. 2011;**27**(2): 205-212

[104] Dwivedi BS, Singh VK, Meena MC, Dey A, Datta SP. Integrated nutrient management for enhancing nitrogen use efficiency. Indian Journal of Fertilisers. 2016;**12**:62-71

[105] Sarkar S. Management practices for enhancing fertiliser use efficiency under rice-wheat cropping system in the Indo-Gangetic plains. Innovare Journal of Agricultural Sciences. 2015;**3**:5-10

[106] Zhang F, Cui Z, Chen X, Ju X, Shen J, Chen Q, et al. 1 integrated nutrient management for food security and environmental quality in China. Advances in Agronomy. 2012;**116**:1-40

[107] Yadav MR, Kumar R, Parihar CM, Yadav RK, Jat SL, Ram H, et al. Strategies for improving nitrogen use efficiency: A view. Agricultural Reviews. 2017;**38**(1)

[108] Zheng Y, Han X, Li Y, et al. Effects of mixed controlled release nitrogen fertilizer with rice straw biochar on Rice yield and nitrogen balance in Northeast China. Scientific Reports. 2020;**10**:9452

[109] Shivay YS, Prasad R, Singh S, Sharma SN. Coating of prilled urea with neem (Azadirachtaindica) for efficient nitrogen use in lowland transplanted rice (*Oryza sativa*). Indian Journal of Agronomy. 2001;**46**:453-457

[110] Liu ZX, Sun L, Wu Y, Liu YL. Effect of coated controlled released urea in yield and nitrogen utilization efficiency of rice in cold region. Journal of North Rice. 2009;**39**:10-15 (In Chinese)

[111] Shoji S, Delegado J, Mosier AR, Miura Y. Controlled release fertilizers and nitrification inhibitors to increase nitrogen use efficiency and to conserve air and water quality. Communications in Soil Science and Plant Analysis. 2001; **32**:1051-1070

Chapter 4

Smart Rice Precision Farming Schemes in Sub-Saharan Africa: Process and Architecture

Abraham Ayegba Alfa, John Kolo Alhassan,
Olayemi Mikail Olaniyi and Morufu Olalere

Abstract

Smart farming integrates information, communication, and control technologies in agricultural practices. Recently, crop enterprise management through smart precision farming technologies are antidotes to uncontrollable soil and environmental factors compounded by climate change. Farm production planning utilizes enormous data generated from the field by human agents and IoT devices, but is often unreliable and inaccurate. These cause low yield, high losses, inferior quality of farm produce, overuse or underuse of fertilizers, increased costs, and inefficient farm management. Traditionally, analyzing rice cropping yields is time-inefficient and tasking, which led to quicker IoT adoption. Aside insufficient data sharing infrastructure, data privacy problem is widespread The blockchain technology is useful for verifying the reliability, accuracy, and authenticity of IoT data generated from fields for the production planning. In the future, dynamic systems (smart rice farming) and model-based control systems can be applied to understand the physical process and valuable factors of production. This paper provides a comprehensive state-of-the-art process and architectural survey on impacts of uncontrollable environmental factors, smart precision framework, security and privacy architectures or solutions for improving rice crop production. Again, a new taxonomy is developed to guide researchers, advance the course of rice production, and improve yields across sub-Saharan Africa.

Keywords: Climate change, rice, precision farming, smart farming, data, security, privacy, blockchain, IoT, inaccuracy, unreliability

1. Introduction

The world's global population is nearly 7.8 billion in 2020 and projected to be 9.7 billion in 2050. Accordingly, these persons need additional 70% of food against the present backdrop of diminishing natural resources such as water and land caused by urbanization, climatic changes, soil erosion, and water shortages. The coping mechanisms for the climate change phenomenon and food security can be best tackled with precision agriculture (or smart farming) [1]. Researchers are exploring viable alternative rice farming methods as means to increase production in peatlands because of their potential to be easily converted into arable farmlands. Rice growth has been impacted mostly by acidic soil and water, irrigation, and rice variety adaptation [2].

This approach holds great promise for better returns and optimizing farming efforts including higher profitability, wastage reduction, and higher productivity. Though, it is nearly impossible for farmers to obtain full-grasp of the various variability linked to the environment because of large inherent uncertainties. Smart farming practice offers to provide farmers the opportunity to simplify and automate the process of acquiring and analyzing farm data, which leads to faster, intelligent, and automated decision-making. The spate of climate change globally has further informed the greatest desire of governments to safeguard food security through maintenance and growth of agricultural production. These rapid climate changes in various abiotic factors including rainfall, drought, temperature, flooding, and solar radiation continue to adversely impact rice production at different stages of growth [3]. The information provided by crops can become a veritable tool for modern agricultural practice in the quest for environmental protection, food production sustainability, cost savings, and improved decision-making [4].

In particular, smart farming makes use of the concept of precision farming to raise farm produce through the understanding and knowledge of the variability of a crop field. It carries out the analysis of the environmental factors and distributes the different inputs based on the specificity of the site (or farmland). This holds great promise for better returns and optimizing farming efforts including higher profitability, wastage reduction, and higher productivity. Though, it is nearly impossible for farmers to obtain full-grasp of the various variability linked to the environment because of large inherent uncertainties. The smart farming practice offers to provide farmers the opportunity to simplify and automate the process of acquiring and analyzing farm data, which leads to faster, intelligent, and automated decision-making [5].

Cultivated land use layout adjustment mechanism for crop planting suitability is concerned with the refinement and deepening of land use transformation. This is of great importance in optimizing the allocation of cultivated land resources and ensuring food security. Presently, farmers and stakeholders infrequently take into account the land suitability for crops when using cultivated land, which results in a mismatch between crop distribution and resource conditions including water, heat, and soil, and adversely affects the ecological security and utilization efficiency of cultivated land. Although, land suitability analysis for crops is a function of crop requirement and land characteristics reflected in final decisions conducted through a multi-criteria evaluation approach [6].

Crop simulation models are used to forecast the impacts of climate change on yield levels, and to identify adaptation strategies [7]. Nevertheless, crop quality has been almost neglected in available studies, despite its relevance to the economic and nutritional value of agricultural products. The few available studies [6, 8, 9] on the impacts of climate change on crop quality are aligned to this trend, depicting an overall decline in many cereal crops. The impacts of climate change were highly divesting on rice milling and cooking suitability performed in the main European rice district, Northern Italy, which cuts harvest into about half of entire production areas.

The projected decay in rice quality due to climate change strengthens the need for modeling studies to forecast its trend in the near future, considering the connections with the stability and utility-safety-quality-nutritive components of food security. Indeed, the expected increase in thermal regimes and in the frequency of weather extremes may cause a higher incidence of damaged grains and reduced head rice yield even in rice top producing areas causing a decreased stability of food production

[7]. The largest impact of climate change on milling suitability was due to the high number of quality variables affected by temperature changes [7].

Water management for rice cultivation is considered a major parameter for low rice yields and large yield gaps. A number of technologies has been investigated by AfricaRice and its partners in the quest to identify their strengths and overcome certain water-related problems. While other studies accounted for fewer rice suitability conditions on lowlands such as rice yield, water productivity, migration of iron toxicity, and water use on the farm field [3].

2. Traditional rice farming vs. smart precision rice farming

Rice (*Oryza spp.)* is an imperative staple food in sub-Saharan Africa. The consumption of rice has risen sharply since the early 1960s because of rising per capita consumption, demographic growth, and urbanization. Regardless of the swelling rate of consumption of rice, local production is within half of demand, and imports usually complement the difference [3]. The obvious shortfalls in rice production across the sub-Saharan Africa are due to concentration of farming activities in smaller rice production areas, and poor yield [3]. The rainfed rice system viability will be impacted by temperature changes and rainfall instability caused by climate changes. Aside from this, other rice cropping systems have emanated in recent times including rainfed upland, rainfed lowland, irrigated lowlands, deep water, and mangrove, which irrigated stand out as most promising [8].

Consequently, the demand and consumption of rice continue to increase significantly as the population grows. In fact, a lot of imported rice is occasioned by low rice production in several nations of Africa in attempt to provide food security. These have further amplified the need to increase rice farming in existing and new lands in order to effectively deploy scarce water resources and nutrients, which serve as limiting factors in both upland and lowland rice growing areas as an important staple crop [10].

Rice is notably the second topmost after wheat consumed grain globally. It has low protein but high carbohydrate composition. Aside from that, it is a rich energy meal for people, which has spiked the demand especially in developing economies [11, 12]. Three most common rice varieties are the lowland Thai species (such as Pathumthani 1, RD57, and RD41), which can be cultivated through transplanting and dry direct seedlings. These are well-known to survive in alternative wetting and drying water supply regime [13]. It needs a minimum of 120 days from time of seedling to harvesting. Thailand, the Philippines, Vietnam, China, and Bangladesh are the world's top producers of rice, which contribute significantly to their Gross Domestic Product (GDP), and economic comparative advantage. Aside from the traditional ways of conserving water and maintenance of soil fertility with adequate mineral content (such as Potassium, Phosphorus, and Nitrogen), there is slow adoption of technologies. The low yields of rice continue to worry farmers and researchers, which is attributed to poor soil fertility, adverse weather, water scarcity, pest attacks, and uncontrollable environmental factors. There is the reluctance of farmers to utilize modern technologies due to uncertainties about safety/privacy and perceived influence on farming yields or outcomes [13, 14]. In sub-Saharan Africa, the lowland rice farming requires keeping the concentrations of phosphorus and water supply at optimal levels for increased yields, which are problematic in meeting the demands for the commodity [10]. The suitability of fields for rice production is associated with socioeconomic and natural factors; thereby impacting the yields.

Climate, topography, and soil indicators characterize the natural conditions of crop planting, which the prerequisite factors are affecting crop suitability. The indicators and criteria for evaluating suitability of land for cropping can be classified into restrictive and nonrestrictive indicators. The restrictive indicators have specific criteria for different crops with the most suitable range, maximum threshold, and minimum threshold, while the nonrestrictive indicators have no specific suitability standards, which need to be further defined. In the case of rice cropping, climatic, terrain, soil, and management factors were noted as impacting the suitability of rice cropping determined by means of [6]. Rice is majorly grown in three environments: rainfed upland, rainfed lowlands, and irrigated lowlands. In all rice-growing environments, the yield gap (that is, the difference between the potential yield in irrigated lowland or water-limited yield in rainfed lowland and upland, and the actual yield obtained by farmers) is largely caused by a wide range of constraints associated with water. Therefore, water management for rice cultivation is impacted by water-related constraints including drought, flooding, iron toxicity, and soil salinity. These required preconditions to achieve higher yield and quality against uncontrolled production factors such as continuous flooding [3].

The theory of precision farming (or precision agriculture) relies on the observing, measuring, and responding to the variability in crop fields or parts of livestock management [5, 10]. It entails applying inputs when and where needed. The precision farming is considered to be the third phase of the revolution in agriculture only trailing behind mechanization and the green revolutions. The main imports of this concept are to increase and maintain the sustainability of crop production and animal rearing through the use of IoT, Artificial Intelligence (AI), robotics, and big data [4]. Smart farming implies the adoption of information and communication technology (ICT) in agriculture [15]. Often, this is referred to as data-driven agriculture in which robotic solutions are incorporated into artificial intelligence techniques for the purpose of attaining the sustainable agriculture of the future [4]. The focus on linking IoT within the sensor networking system is emerging in smart and precision farming. Reason is that IoT is capable of acquiring information (such as the moisture rate), and broadcasting to the user in wireless setups.

The traditional methods of farming are incapable of satisfying the feeding needs of people across the globe. So, IoT-driven smart farming becomes an unavoidable mode of agricultural information management. Smart farming can provide remote monitoring and control of a farm equipment to improve the productivity and quality and prevent disasters. Nevertheless, certain technical hurdles need to be addressed in terms of smart farming including lack of data sharing infrastructure because no mechanism is available to provide privacy protection in sharing sensitive agriculture data leading to irregular monitoring of farming systems [16]. Again, proper management of farm resources such as water, fertilizers, and seed quality has direct influence on the produce. Universally, majority of farms are still operating at small-scale levels covering a number of acres due to over-reliance on customary farming practices. The outcomes are often low yield, high crop losses, overuse or inadequate use of fertilizers, and soil spoilage [17].

The focus on linking IoT devices within the sensor networking system is emerging in smart and precision farming. Reason is that IoT devices are capable of acquiring information (such as the moisture rate) and broadcasting to the user in wireless setups. Therefore, farmers become significantly aware of the state of the farm field remotely in order to perform computation, collect data, and link with other nodes throughout the region of interest [18]. In particular, majority of IoT devices are susceptible to compromises and hacks, which render their ensuing data unreliable for productive ventures. Again, IoT devices have limited computational, memory, and network aptitude. These make them more prone to attacks against endpoint devices, such as tablets, smartphones, or PCs [19]. The IoT is targeted at improving human life and releasing vast economic value. But, insufficient data security and trust in contemporary IoT have stood in way of full exploitation and adoption [20].

The notion behind smart farming system is to intellectualize management of crop enterprises involved in deployment of precision farming technologies. This paradigm hampers the knowledge bases and multi-agent technologies to advance coordinated decisions on planning, distribution, control, and optimization of enterprise resources on real-time basis [21]. In particular, farm production planning relies heavily on enormous data generated from the field such as the soil infiltration characteristics by human agents and IoT devices, but, is often manipulated and tempered, rendering them unreliable and inaccurate for appropriate modeling procedures [22].

Traditional approaches for crop (such as rice) investigation and analysis of soils are time-inefficient and tasking, which led to the quest to deploy IoT and secure data for providing remote soil and environmental parameters monitoring. Smart farming is facing insufficient data sharing infrastructure because of poor mechanisms to broadcast sensitive agriculture data in a privacy-protected form [16].

Again, the future trends and behaviors of dynamic systems such as rice farming and model-based controls are applied to understand the physical process and valuable factors of production [23]. Therefore, prediction tasks are conducted by smart and precision-based systems, which rely heavily on data powered by IoT devices and things.

IoT-based systems are highly susceptible to attacks, that is, manipulation and tampering of data on unsecured over long-distance wireless network coverage. Likewise, the privacy of device owners is not guaranteed due to autonomous data sharing across diverse entities including cloud, fog, and Internet [24]. Whenever threats of any kind can be successfully launched on wireless data transmission channels or storage infrastructures, privacy leakage will happen more quickly from insider or outsider agents [25].

IoT is the main motivation for a paradigm leap to a truly connected world in which everyday objects become interconnected, capable of communicating directly with each other, and capable of collectively offering smart services [26]. Therefore, in several of these applications, the data collected by IoT is sensitive and must be kept private and secure [27]. Several issues have continued to limit the widespread utilization of IoT as reported by numerous scholars as follows: According to Yang *et al.* [28], IoT device safety remained the topmost challenge due to attacks such as Mirai's Botnets of Things (DDoS). There is no adequate data protection mechanism at present for IoTs. The number of device types (homogeneity and heterogeneity) affects security of IoT. The devices are low-resource type and relatively expansive in quantity.

The Blockchain technology has the potential of providing privacy preservation in IoT-based smart farming for its sensitive data generation and communication in decentralized scenarios [29], as well as the overall improvement of farming practice or produce. The usefulness of blockchain technology in providing distributed things security services including confidentiality, privacy, provenance, authentication, and integrity was highlighted by Salman *et al.* [30]. Accordingly, authentication and confidentiality are attainable through the public key cryptography (that is, encryption and signature approaches). There is a need to investigate the various blockchain approaches in large-scale and real-world situations for performance assessments,

Farming types	Traditional rice farming	Smart precision rice farming
Scale	Subsistence or small	Large
Strengths	Natural ways of conserving water and maintenance of soil fertility. Natural conditions reliant.	Controllable and management of production factors. High deployment of technology. Data-driven farming.
Weaknesses	Scarce water resources and nutrients. Low yields. Uncontrollable environmental factors. Large uncertainties.	Low adoption of technology. Limited computational, memory, and network aptitude. IoT or data are vulnerable to privacy compromises.
Method(s)	Rainfed lowlands	Rainfed uplands and irrigated lowlands.
Human intervention	High	Low
Quality of Yield	Low	High
Research funding prospect(s)	Low	High
Modeling procedure	Inaccurate and time-consuming.	Accurate and faster.
Generation	Pre-Internet era	Post-Internet era
Key terms	Basic farm tools, and mechanization.	Data, technology, computer-aided decision-making, and IoT.

Table 1.
Distinction between traditional and smart precision rice farming.

especially in smart precision farming. The key distinctions between the traditional rice farming and smart precision rice farming are depicted in **Table 1**.

3. Impacts of farm parameterization on rice production

In agriculture, the fertility of soil refers to the capacity of soil to support crop growth by supplying required soil nutrients, as well as suitable biochemical, and physical properties as a growing environment for plants. Particularly, majority of the available research investigated a few indicators, such as rice yield, water use, and water productivity at the field level. There are sparingly limited works on the cost–benefit of water management technologies, enabling conditions, and business models for their large-scale adoption, with regard to their impact on farmers' livelihoods (women and youths) [3]. Besides, water management design for crop diversification, landscape-level water management, and iron toxicity mitigation, particularly in lowlands have received strained attention. More so, climate, topography, and soil indicators are natural conditions and prerequisites for determining rice cropping suitability.

Rice consumption has steadily increased in sub-Saharan Africa while domestic rice production hardly meets the demand. There are five rice cropping systems in West Africa including rainfed upland, rainfed lowlands, irrigated lowlands, Deepwater, and Mangrove swamps [8]. But, the irrigated rice systems hold promise for the future in numerous ways: Firstly, the average rice yield in irrigated lowland is better than yields in rainfed lowland, and rainfed upland. Secondly, due to temperature changes, rainfall variability, and expected future climate change impacts in rainfed rice systems, improvements in farmers' adaptive capacity due to the expansion of irrigation facilities may reduce rice production losses. Although, irrigated rice holds tremendous potential in fulfilling many West African countries' agendas of becoming rice self-sufficient, geospatial analysis to assess potentially irrigable land is often unexplored.

In particular, realizing the rice planting environmental patterns are capable of enhancing natural and human-environmental resources, which support sustainability in the long run. Rice yields are impacted by soil, topography, climate, farmland management, agricultural mechanization, and many other artificial environmental factors. These are required for planning and improving the high-precision systems for forecasting rice yields [9]. The new challenges and expectations of subsequent research efforts can be in the following ways:

i. The development of localized solution to solve rice production and food sustainability in horrendous climatic conditions in sub-Saharan Africa.

ii. The use of precision technique for environmental parameters optimization and soil characterization in smart rice farming.

iii. The privacy of data stored in the cloud is improved with blockchain technology.

iv. The privacy problem of blockchain technology can be demystified with lightweight cryptographic scheme in order to improve reliability and trust of cloud data storage.

In particular, farm production planning relies heavily on enormous data generated from the field, such as the soil infiltration characteristics by human agents and IoT devices, but, is often manipulated, and tempered, rendering them unreliable and inaccurate for appropriate modeling procedures [22].

According to Banerjee, Lee, & Choo [31], there is lack of publicly obtainable dataset from IoT devices deployed for farming practices (such as rice) because of non-existing data sharing standards for enforcing integrity in order to assist researchers.

4. Existing frameworks of smart farming systems

The spate of climate change globally puts burden on governments in safeguarding food security through increased agricultural production. These rapid climate changes radiations continue to adversely impact rice production throughout growing stages based on different abiotic factors, such as rainfall, drought, temperature, flooding, and solar [32]. Precision farming is based on the observing, measuring, and responding to the variability in crop fields for strategic management [1, 5, 33]. It entails applying inputs when and where needed as means of sustaining crop production. The basic problems facing agricultural production are crop selection, decision-making, and supporting systems for better crop yield. Agriculture prediction is associated with natural parameters including temperature, soil fertility, water quality, water volume, season, and crop prices. A smart crop monitoring and tracking framework composed of sensors, mobile applications, IoT cameras, and big data analytics were proposed by [34]. While the hardware component consists of an Arduino UNO, a variety of sensors, and a Wi-Fi module without privacy preservation strategy.

An information-based management cycle for advanced agriculture was proposed by [4]. The framework contains crop sensing objects, platform, data warehouse, decision system, and actuation. There is no consideration for privacy of collected data throughout the cycle. Smart farming drive in part targeted the wireless sensor network in the attempt to deal with the developing problems of global warming. Sensors have potentials to effectively collect the farming environmental parameters for making smarter and faster decisions using computer-based solutions. Crop yield depends greatly on the ability of farmers to make faster and accurate decisions [18].

One key problem facing society at present is the privacy and security of available data. There are issues surrounding collection, storage, integration, and transformation of data to support the needs of agri-food (Mushroom production) due to widespread deployment of evolving technologies. Apart from competitive advantage and increased production, environment variables need to be regulated because of their direct relationship to production control system. One form of farming was understudied by [35], which is an extension mushroom farm distributed process control system using IoT and Blockchain to facilitate distributed data collected on environmental indicators to supplement the production control system for mushroom farming. Again, there is improvement in farm management information system available to managers of farms. But, precision farming is expected to have comprehensive decision support system from its subsystems with privacy in mind.

The evolving ecosystem control system and technology are still incapable of providing the required intelligence for rural farms. Inadequacy of precise information and communication negatively impacts the farm production management, which was overtaken by IoT rather than conventional farming practice [36]. This led to introduction of framework for constructing fresh technology and application for IoT-based farming. They enhance farming processes, monitoring, and welfare of farms and farmers with unrestricted access to data in real time. Of course, there is greater leap toward food sustainability by means of connected objects. The potential of IoT to advance farming sector and food sustainability for populace was undertaken by [22]. Smart farming is a potent coping mechanism for extreme weather and climatic conditions as well as massive demands for farm produce. This enhances productivity and grows cleaner food to cater to exploding population. In the complete framework for the smart farming system, processes and requirements left out the most important component, that is, privacy of data generated from IoT devices or transmitted across vulnerable public channels.

In [24], the promise of fog computing and wireless networks to interconnect farm bases distributed in rural setups became feasible. To this end, a scalable network structure for controlling and monitoring farms in rural areas evolved with cross-layered channel access and routing schemes. There are noticeable improvements in latency, sensing, and actuating but, chances of tampering and manipulating data are highly probable. Specifically, the gaiasense™ smart farming framework is composed of IoT devices (or GAIAtrons), cloud computing services (or GAIA Cloud), and smart farming (SF) advisory services (or GAIA SmartFarm) [37]. However, privacy of data was not considered.

The coping mechanism for the climate change phenomenon and food security can be best tackled with precision and smart farming [1]. Typically, smart precision farming system frameworks are composed of IoT data acquisition, IoT data storage, data sharing, and IoT data analytics as depicted in **Figure 1**.

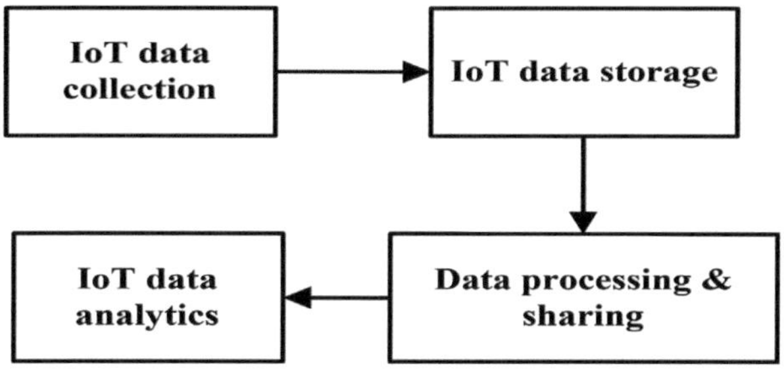

Figure 1.
The typical framework of smart precision farming system.

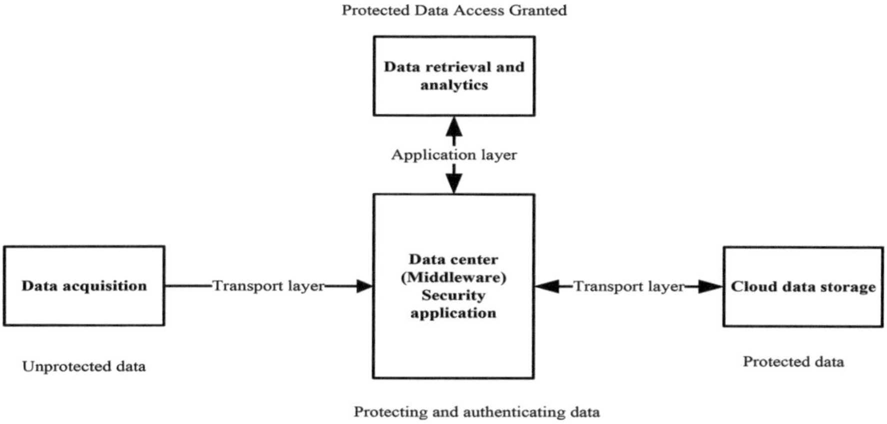

Figure 2.
The data privacy protection strategy for smart rice farming scheme.

The issue of privacy of data is topmost when IoT and blockchain technologies are deployed for smart precision farming practices. Therefore, this paper proposes a new framework leveraging the shortfall of aforementioned frameworks to develop a smart precision system for determining the suitability of environmental and soils for rice cropping. On the other hand, the complete model of data and transactions protection strategy proposed by this paper for smart precision rice farming system is shown in **Figure 2**.

In **Figure 2**, the data privacy protection strategy commences at the IoT device **transport layer** by offloading the protecting operation based on **lightweight cryptosystem** to develop at the middleware (or data center). Then, the protected data is broadcast to the blockchain cloud data storage at the **middleware layer** transport layer. The protected data can be accessed through **authenticating operations** at the application layer.

5. Future architecture of smart precision Rice farming systems

The traditional methods of farming are incapable of satisfying the feeding needs of people across the globe. So, the IoT-driven smart farming becomes an unavoidable mode of agricultural information management. But, insufficient data security and trust in contemporary IoT have stood in way of full exploitation and adoption [38].

The strengths of precision models depend on available, reliable data about farmland, soil characterization, infiltration characteristics, and chemical nutrients, which are often difficult to attain due to distributed and complex nature of harvesting and analyzing data in developing countries. The structure of the proposed smart rice precision is composed of rice farm field parameters composed of the sensors, embedded system, and the cloud storage.

The research process design is subdivided into four main steps discussed as follows:

Phase 1 is data capturing: This is entirely the requirements and problems specification for the proposed scalable solutions.

Phase 2 is the formation of a data encryption model: The phase includes formulation lightweight cryptosystem to generate sizes of keys and block-ciphers, which protect data against privacy compromises in smart farming scenario.

Phase 3 is the formation of data processing model: The phase includes the use of data mining, machine learning, and forecasting estimators for understanding the variability of data captured.

Phase 4 is the development: This involves the implementation of the proposed permissioned blockchain-based smart precision rice farming.

Phase 5 is the evaluation: It includes testing the proposed smart rice precision system with selected dataset to ascertain the performance.

The structure of the proposed smart rice precision is composed of rice farm field parameters composed of the sensors, embedded system, and Blockchain security in the cloud storage as shown in **Figure 3**.

In **Figure 3**, the series of events for the proposed secure smart rice farming data are indicated with letters A-E due to the distinct roles played throughout the data lifecycle.

Event A: This entails deployment of field sensors and GPS to capture distinct parameters required for rice farming and production broadcast across embedded system unit.

Event B: The data are sent from embedded system unit and received at the Blockchain. Then, the proposed lightweight cryptosystem is to be used for data key generation and hashes.

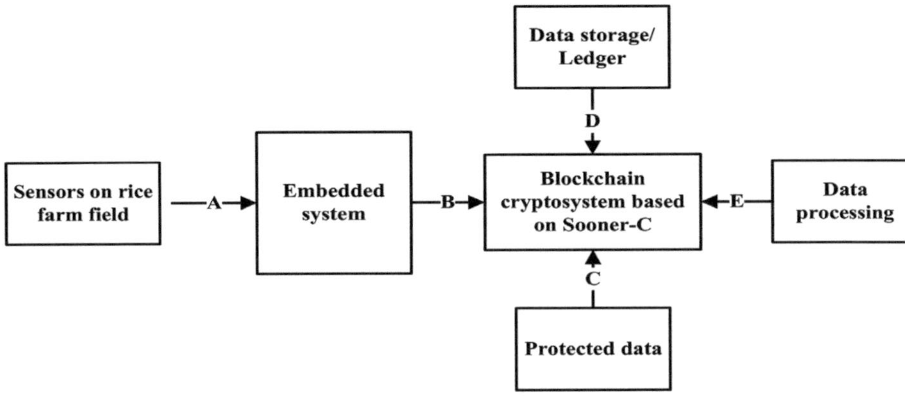

Figure 3.
The events modeling of smart precision rice farming system.

Event C: The encrypted data are stored on the distributed ledger provided by Permissioned Blockchain (PBC).

Event D: The data are protected using the lightweight scheme for effective processing and computation in order to enforce privacy.

Event E: The data processing can be performed on protected data using proof-of-work. This way the security and privacy of data are preserved.

The research concerning smart precision rice farming can border on the following aspects including data acquisition, data analytics, climatic factors, soil factors and water conservations, environmental factors, and topology/toposequence as illustrated in **Figure 4**.

Figure 4 provides the direct relationships between the rice production and uncontrollable environmental factors debacle in sub-Saharan Africa. The process of overcoming the climate change problems requires that, the challenges of rice field dataset and the inaccurate decision-making about parameters required for rice production. In particular, IoT data controls, governance, and privacy/security continue to bedevil their progress. The method with reoccurring assurances to withstand the shocks of climate change phenomenon is the smart farming, which utilizes IoT (sensors and actuators) to harvest dataset from the field to the point of usage. Precision farming relies on smart datasets to project parameter dynamics at certain points in time especially climate, soil, and environmental factors. The combination of the two technologies ignites the potential to increase rice production despite large instability in climatic conditions around the sub-Saharan Africa. There is a need to the redesign data transmission channels and network connectivity between IoT devices and the cloud storage infrastructure for data privacy and effective exchanges.

The key solutions for improving the fight against climate dynamics are appropriate data warehousing from numerous IoT placed on the rice fields; secure and privacy-protected data sharing among stakeholders, data providence by means of blockchain technology; and the use of high-performance AI techniques including machine learning, deep learning, and natural processing. In this way, the process of arriving at quicker, safe, and accurate, decision-making can be automated concerning suitability of lands for rice crop production as against the less-effective manual approaches. This taxonomy adds to the knowledge of the relevant players in rice production and climate controls by minimizing wastage and parameter forecasting.

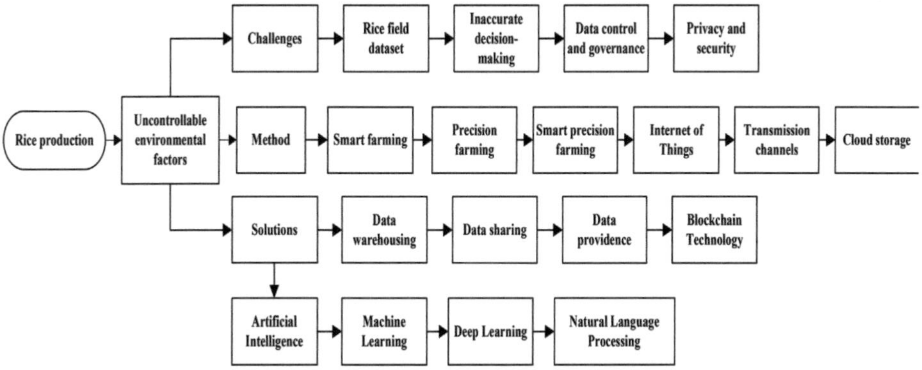

Figure 4.
Taxonomy of smart precision farming for rice production.

6. Conclusion and recommendations

Population growth is exponential, yet land is a limited and precious resource on the planet. Thus, it is critical to utilize that fixed resource sustainably in order to meet rising food demand without damaging the land with improper land practices. To reduce the undesirable impact of environmental features on agricultural productivity and on land, land suitability analysis is the first and foremost step for environmental management and sustainable agriculture. Land suitability assessments not only aid in increasing crop yields but also in maintaining healthy soil conditions for bountiful output. Rice planting suitability depends on slope and geomorphological formations. This paper developed a framework for smart precision rice farming system that utilizes sensors, blockchain technology, and AI approaches for determining the environmental factors, soil fertility, and suitability of farmlands for rice cultivation in semi-humid climatic zone of sub-Saharan Africa. The proposed framework attempts to improve the renowned syn. Jenks (natural breaks) method for seasonal farming practices. The devastating effects of climate change on rice production will continue to worsen with attendant negative implications on productivity and rice farming practices in sub-Saharan Africa. Though, the proposed smart precision rice farming systems can effectively deal with the climate change debacle in the short and long runs.

The key recommendations from this paper include development of full-fledged smart precision farming systems for the local farming scenarios in order to reduce resource wastages and environmental hazards while increasing rice crop health; development of effective lightweight privacy-preserving schemes in smart data centers for data governance and providence; the adoption of viral and high-performance analytical mechanisms to support faster and accurate decision-making process on rice production planning and parameters estimations. The target audience for this paper is primarily the research community, farmers, and investors by providing them with supportive tools, knowledge, and decision-making mechanisms on process and architecture of smart precision rice farming schemes. The outcomes shall be helpful to government donors, regulatory agencies, and international organizations on the need to improve and adopt responsible technology to combat hunger, low rice production, resource wastages, resource conservation, and climate problems through smart systems and AI.

DOI: http://dx.doi.org/10.5772/intechopen.107541

Author details

Abraham Ayegba Alfa[1,2]*, John Kolo Alhassan[2], Olayemi Mikail Olaniyi[2] and Morufu Olalere[2]

1 Confluence University of Science and Technology, Osara, Nigeria

2 Federal University of Technology, Minna, Nigeria

*Address all correspondence to: mikail.olaniyi@futminna.edu.ng

References

[1] Alfred R, Obit JH, Chin CPY, Haviluddin H, Lim Y. Towards paddy rice smart farming: A review on big data, machine learning, and rice production tasks. IEEE Access. 2021;**9**:50358-50380

[2] Fawzi NI, Qurani IZ. Lesson learned from the development of sustainable rice farming in peatland. IOP Conference Series in Earth Environmental Science. 2021;**752**:1

[3] Dossou-yovo ER, Prasad K, Akpoti K, Danvi A, Duku C, Zwart SJ. Field crops research thirty years of water management research for rice in sub-Saharan Africa: Achievement and perspectives. Food Crops Research. 2022;**283**:108548

[4] Saiz-Rubio V, Rovira-Más F. From smart farming towards agriculture 5.0: A review on crop data management. Agronomy. 2020;**10**:2

[5] Bala JA, Folorunso TA, Olaniyi OM, Daniya E. Design of an Agribusiness Innovative and Autonomous Robot System for Chemical Weed Control for Staple Food Crops Production in Sub-Saharan Africa. In: Che FN, Strang KD, Vajjhala NR, editors. The Advances in Business Information Systems and Analytics (ABISA). Hershey, PA: IGI Global; 2021. pp. 234-262

[6] Song G, Zhang H. Cultivated land use layout adjustment based on crop planting suitability: A case study of typical counties. Land. 2021;**10**(107):1-19

[7] Cappelli GA, Bregaglio S. Model-based evaluation of climate change impacts on rice grain quality in the main European rice district. Food Energy. 2021;**10**(4):1-22

[8] Akpoti K, Dossou-yovo ER, Zwart SJ, Kiepe P. The potential for expansion of irrigated rice under alternate wetting and drying in Burkina Faso. Agricultural Water Management. 2021;**247**:106

[9] Li D, Liang J, Wang X, Wu S, Xie X, Lu J. Rice yield simulation and planting suitability environment pattern recognition at a fine scale. International Journal of Geo-Information. 2021;**10**(612):1-19

[10] Bauw P et al. Field crops research combining phosphorus placement and water saving technologies enhances rice production in phosphorus-deficient lowlands. Food Crop Research. 2019;**236**(March):177-189

[11] Thakur D, Kumar Y, Kumar A, Singh PK. Applicability of wireless sensor networks in precision agriculture: A review. Wireless Communication. 2019;**107**(1):471-512

[12] Özkan B, Dengiz O, Turan İD. Site suitability assessment and mapping for rice cultivation using multi-criteria decision analysis based on fuzzy - AHP and TOPSIS approaches under semihumid ecological condition in delta plain. Paddy Water Environment. 2019;**17**(4):665-676

[13] Ullah H, Rahimi AZ, Datta A. Growth and yield of lowland rice as influenced by potassium application and cultivation method under alternate wetting and drying water regime. Journal of Plant Nutrition. 2019;**42**(13):1529-1542

[14] Zhuang Y, Zhang L, Li S, Liu H, Zhai L, Zhou F, et al. E ff ects and potential of water-saving irrigation for rice production in China. [internet].

Agricultural Water Management. 2019;**217**:374-382

[15] Bacco M, Barsocchi P, Ferro E, Gotta A, Ruggeri M, Bacco M, et al. The digitisation of agriculture: A survey of research activities on smart farming. Array. 2019;**3**:1000

[16] Patil AS, Tama BA, Park Y, Rhee KH. A Framework for Blockchain Based Secure Smart Green House Farming. Singapore: Springer Verlag; 2018. pp. 1162-1167

[17] Na A, Isaac W, Varshney S, Khan E. An IoT based system for remote monitoring of soil characteristics. In: 2016 International Conference on Information Technology (InCITe)- The Next Generation IT Summit on the Theme - Internet of Things: Connect your Worlds. Noida, India: Institute of Electrical and Electronics Engineers; 2016. pp. 316-320

[18] Surai S, Kundu R, Ghosh R, Bid G. An IoT based smart agriculture system with soil moisture sensor. Journal of Innovation Research. 2018;**1**(1):39-42

[19] Ahmad M, Salah K. IoT security: Review, blockchain solutions, and open challenges. Future Generation Computer System. 2018;**82**:395-411

[20] Wang X, Zha X, Ni W, Ping R, Guo YJ, Niu X. Survey on blockchain for internet of things. Computer Communication. 2018;**136**:10-29

[21] Skobeleva PO, Simonovac EV, Smirnova SV, Budaeve DS, Voshchuke GY, Morokovd AL. Development of a Knowledge Base in the "smart farming" system for agricultural Enterprise management. Procedia Computer Science. 2019;**150**:154-161

[22] Mat I, Kassim MRM, Harun AN, Yusoff IM. Smart agriculture using internet of things. In: 2018 IEEE Conference on Open Systems (ICOS). Langkawi, Malaysia; 2019. pp. 54-59

[23] Yu W, Pacheco M. Impact of random weights on nonlinear system identification using convolutional neural networks. Information Science. 2018;**477**:1-14

[24] Ahmed N, De D, Hussain I. Internet of things (IoT) for smart precision agriculture and farming in rural areas. IEEE Internet of Things Journal. 2018;**5**(6):4890-4899

[25] Qiu M, Gai K, Zhao H, Liu M. Privacy-preserving smart data storage for financial industry in cloud computing. Concurrnce Computer Programming Expertise. 2017;**30**:1-10

[26] Georgakopoulos D, Jayaraman PP. Internet of things: From internet scale sensing to smart services. Computing. 2016;**98**(10):1041-1058

[27] Kouicem DE, Bouabdallah A, Lakhlef H. Internet of things security: A top-down survey. Computer Networks. 2018;**141**:1-24

[28] Yang Y, Yang Y, Chen J, Liu M. Application of blockchain in internet of things. In: Lecture Notes in Computer Science (including subseries Lecture Notes in Artificial Intelligence and Lecture Notes in Bioinformatics). Switzerland: Springer Verlag; 2018. pp. 73-82

[29] Ejaz W, Anpalagan A. Blockchain Technology for Security and Privacy in Internet of Things. Switzerland: Springer, Charm; 2018. pp. 47-55

[30] Salman T, Zolanvari M, Erbad A, Jain R, Samaka M. Security services using blockchains : A state of the art survey. IEEE Communication Surveys and Tutorials. 2019;**21**(1):858-880

[31] Banerjee M, Lee J, Choo KKR. A blockchain future for internet of things security: A position paper. Digit Commununication Networks. 2018

[32] Hussain S, Huang J, Huang J, Ahmad S, Nanda S, Anwar S, et al. Rice production under climate change: adaptations and mitigating strategies. In: Environment, Climate, Plant and Vegetation Growth. Switzerland: Springer; 2020. pp. 659-686

[33] Mizik T. How can precision farming work on a small scale? A systematic literature review. Precision Agriculture. 2022;**4**(3):149-160

[34] Phasinam K, Kassanuk T, Shabaz M. Applicability of internet of things in smart farming. Journal of Food Quality. 2022;**2022**:1-7

[35] Branco F, Moreira F, Martins J, Au-Yong-Oliveira M, Gonçalves R. Conceptual approach for an extension to a mushroom farm distributed process control system: IoT and blockchain. In: Rocha Á et al., editors. WorldCIST'19 2019, AISC 930. Switzerland: Springer; 2019. pp. 738-747

[36] Hari Ram VV, Vishal H, Dhanalakshmi S, Meenakshi Vidya P. Regulation of water in agriculture field using Internet of Things. In: 2015 IEEE Technological Innovation in ICT for Agriculture and Rural Development (TIAR). Chennai, India; 2015. pp. 112-115

[37] Kalatzis N, Marianos N, Chatzipapadopoulos F. IoT and data interoperability in agriculture: A case study on the gaiasenseTM smart farming solution. Global IoT Summit. 2019;**2019**:1-6

[38] Wang X, Zha X, Ni W, Liu RP, Guo YJ, Niu X, et al. Survey on blockchain for Internet of Things. Computer Communications. 2019;**136**:10-29

Chapter 5

The Potential Health Benefits of Brown Rice

Shaw Watanabe

Abstract

In many countries, rice contributes to better health by supplying dietary energy, proteins, fat, and various micronutrients. Many different rice species are cultivated in Japan and other rice-producing countries, in which we expect some varieties to prevent many diseases. In particular, the health effects of brown rice are apparent. In particular, rice bran ingredients accumulated evidence about their physiological and pharmacological activity. The Japanese diet has become a world heritage and famous worldwide, but knowledge about the benefits of rice eating is limited. Here, we would like to focus on the benefits of eating brown rice and recently developed low-protein fermented brown rice (LPFG) to improve the gut-kidney axis's negative spiral in kidney disease patients. Other potential benefits of brown rice are the suppression of dementia and celiac disease. The category of "medical rice" represents the health effects of rice eating.

Keywords: Brown rice, functional ingredients, gut microbiota, short-chain fatty acids, medical rice, low protein fermented genmai, GENKI study

1. Introduction

Approximately 70 percent of the world's population eats rice as a staple food, especially in areas of the Asia-Pacific region [1]. Rice can supply energy, proteins, and fat, which accounts for more than half of the diet in southeastern countries [2]. In the 19th century, traditional Japanese meals were of unpolished brown rice, miso soup, and dishes of soybean products, vegetables, potatoes, and other roots. In the Meiji era, polished white rice became popular, and beriberi increased in the nation. After World War II, western life habits introduced white bread, meat, eggs, and dairy products, which became the significant foods composing main and side dishes. Consequently, obesity and other lifestyle-related chronic diseases have increased [3].

Another traditional way of eating in Japan is macrobiotics. Yukikazu Sakurazawa (George Ohsawa) made the foundation and emphasized genmai (brown rice) and whole foods as the central dogma of the macrobiotic diet [4]. Macrobiotic meals, such as seasonal vegetables, beans, and sea vegetables with brown rice, are plant-based and contain functional factors in addition to ordinary nutrients.

In this regard, the nutritional effects of rice should be more studied, especially the integrated composition of functional ingredients [5, 6]. We cultivate more than 200 different rice species in Japan. We can use some varieties for dietary therapy. For example, high-amylose hard rice is available for diabetic patients [7]; low-protein rice for patients with chronic renal disease [8, 9]; GABA-rich large germ rice for mental health [10]; and rice with high antioxidant activity may be effective for the prevention of cancer and other diseases [11, 12].

The health effects of brown rice are traditionally well known. We proposed to summarize the above rice in the new concept, "medical rice." Examples are medical rice for diabetes, medical rice for chronic kidney disease (CKD), medical rice for mental health, medical rice for cancer prevention, etc. [13].

In response to the enormous increase in medical costs in many countries, achieving healthy longevity by changes in dietary habits is mandatory. The proper food labeling of evidence-based medical rice can treat people who want to improve their health [14].

Human data are increasing, so we believe it is time to implement the concept of medical rice for disease prevention and treatment [15].

2. Functional ingredients in rice

2.1 Functional ingredients in rice

Rice can supply energy, proteins, fat, vitamins, and minerals [5, 6, 15]. Red rice and black rice have rich dietary fiber, but it is not popular because of their taste (**Table 1**).

Whole brown rice contains rich vitamins, minerals, dietary fibers, and various functional chemicals compared with white rice. About 8.5 million tons of brown rice are produced annually in Japan, and rice bran makes up about 10% of unprocessed rice by weight and contains 18–22% oil. Rice oil contains up to 5% of unsaponifiable dark oil. The active ingredients in the oil are γ-oryzanol, tocotrienol, GABA, and inositol [16–18].

We can apply rice bran to food, animal feed, and fertilizer but discard most of the bran. Recently, we have paid much attention to rice bran because of the various pharmacological properties of its ingredients [19–21]. We separate into gum, wax, dark oil, and scum at different boiling temperatures, and further extraction yields many chemicals with biological activities.

γ-oryzanol is a potent functional ingredient in the nonsaponifiable fraction of rice bran. γ-oryzanol is bound in 4 chemical forms to ferulic acid and thus belongs to the family of ferulated sterols [20, 21]. The solubility of γ-oryzanol is only 0.06% in water and 0.2% in 20% ethanol. The absorption of γ-oryzanol may not be optimal by oral intake of brown rice, but it is possible to take 300 mg of γ-oryzanol orally from brown rice. Kokumai et al. [22] administered a single oral administration of 300 μmol/kg body weight of rice bran γ-oryzanol to rats and showed that intact γ-oryzanol, along with ferulic acid and ferulic acid conjugates, existed in the blood.

2.2 Nutritional aspect of brown rice

We have studied the macrobiotic practitioners by GENKI study (genmai epidemiology nutrition and kenko innovation) [23]. They consumed more magnesium, iron, vitamin Es, vitamin Bs, and dietary fibers, although their energy intake is less than that of ordinary Japanese. Their blood pressure, body mass index (BMI), and low-density lipoprotein cholesterol levels are low, while blood glucose and glycated hemoglobin (HbA1c) remained within normal levels.

boiled rice 100 g	unit	Brown rice	White Indica rice	White Japonica	White no amylose	Red rice*	Black rice*
Energy	kJ	647	781	663	801	636	634
Energy	kcal	152	184	156	188	150	150
Water	g	60.0	54.0	60.0	52.1	61.3	62.0**
Protein	g	2.8	3.8	2.5	3.5	3.8	3.6
Fat	g	1.0	0.4	0.3	0.5	1.3	1.4
Carbohydrate	g	32.0	37.3	34.6	41.5	28.2	28.2
Fiber	g	1.4	0.4	0.3	(0.4)	3.4	3.3
Soluble	g	0.2	0	0	0	0.2	0.3
Insoluble	g	1.2	0.4	0.3	0.4	2.3	2
Polyalcohol	g	—	—	—	0	0	—
Carbohydrate_all	g	35.6	41.5	37.1	43.9	32.7	32.2
Organic acid	g	—	—	—	—	—	—
ash	g	0.6	0.2	0.1	0.1	0.6	0.6
Na	mg	1	0	1	0	1	Tr
K	mg	95	31	29	28	120	130
Ca	mg	7	2	3	2	5	7
Mg	mg	49	8	7	5	55	55
P	mg	130	41	34	19	150	150
Fe	mg	0.6	0.2	0.1	0.1	0.5	0.4
Zn	mg	0.8	0.8	0.6	0.8	1.0	0.9
Cu	mg	0.12	0.10	0.10	0.11	0.12	0.11
Mn	mg	1.04	0.42	0.35	0.50	1.00	1.95
Se	mg	1	3	1	1	1	2
Cr	mg	0	1	0	0	Tr	1
Mo	mg	34	32	30	48	24	33
Beta carotene	mg	0	(0)	0	(0)	1	8
Vitamin D	mg	(0)	(0)	(0)	(0)	—	—
Tocophrol α	mg	0.5	0	Tr	(Tr)	0.6	0.3
Vitamin K	mg	(0)	(0)	(0)	(0)	—	—
Thiamin	mg	0.16	0.02	0.02	0.03	0.15	0.14
Riboflavin	mg	0.02	Tr	0.01	0.01	0.02	0.04
Niacin	mg	2.9	0.3	0.2	0.2	2.8	3.0
Niacin eq.	mg	3.6	1.3	0.8	1.0	3.4	3.6
Vitamin B6	mg	0.21	0.02	0.02	(0.02)	0.19	0.18
Vitamin B12	μg	(0)	(0)	(0)	(0)	—	—
Folate	μg	10	6	3	(4)	9	19
Pantothenic acid	mg	0.65	0.24	0.25	(0.30)	0.47	0.40

boiled rice 100 g	unit	Brown rice	White Indica rice	White Japonica	White no amylose	Red rice*	Black rice*
Biotin	µg	2.5	0.5	0.5	(0.5)	2.8	2.7
Vitamin C	mg	(0)	(0)	(0)	(0)	—	—
NaCl eq.	mg	0	0	0	0	0	0

**Polyphenol 0.2 g.*
***DF by AOAO2011.25 method.*
Insoluble dietary fiber is rich in brown rice, especially red and black rice. Red and black rice contain about 0.2 g of polyphenol, so the taste is not good. No amylose rice is used for "mochi" (sticky rice cake).

Table 1.
Nutrients in 100 g boiled rice.

The macrobiotic dietary habit of eating brown rice contributed to their healthy state [24–29]. The consumption of small fish provided vitamin B12, eicosapentaenoic acid (EPA), and docosahexaenoic acid (DHA). Genmai-shoku (diet) provides enough energy, fat, protein, and several times more minerals and vitamins than required [26, 30].

In addition to the functional effects of ingredients, chewing frequency influences brain function in genmai eaters [31]. In Japan, fast foods with soft textures have recently become popular with a younger generation, and the oral biting frequency has decreased in the younger proportion. Boiled brown rice increases the chewing number of times than meat or fish dishes, and it was only 800 times per American dish compared to the 30,000 times by the genmai diet. Longer eating time prevents fast eating, which used to lead to obesity and gives relaxation to solve stress.

So, brown rice could be a core of "medical rice for health" strategy. In addition, organic rice is free from arsenic and other toxic chemicals absorbed from fertilizers and insecticides [5, 31].

2.3 Functional ingredients in rice bran

2.3.1 Lipophilic ingredients

Phenolic compounds are rice's primary antioxidant and radical scavenging ingredients [32–35]. In brown rice, the three most abundant ones are 6'-O-feruloylsucrose, 6'-Osinapoylsucrose, and ferulic acid. Their representative concentrations of 1.09, 0.42, and 0.33 mg/100 g rice flour represent 84.0% by weight of the total amount of soluble phenolic compounds (2.19 mg/100 g brown rice flour). Polished rice contains only 0.28 mg of phenolic compounds/100 g of rice flour.

Lipophilic ingredients can decrease total and LDL cholesterol; triacylglycerol and ApoB. They increase HDL cholesterol and inhibit platelet aggregation. Ferulic, caffeic, sinapinic, p-coumaric, vanillic, protocatechuic, syringic, hydroxybenzoic, and chlorogenic acid are also present, but the dose is scanty.

Tocopherol, triacylglycerol, and ApoB are lipid-soluble antioxidants that prevent cardiovascular diseases and cancer. Squalene, an isoprenoid compound structurally similar to beta-carotene, is an intermediate metabolite in synthesizing cholesterol. In humans, about 60 percent of dietary squalene is absorbed. Lipophylic antioxidant is rich in genmai [36].

2.3.2 Water-soluble ingredients

Inositol and phytic acid are water-soluble ingredients like GABA [37–39]. Magnesium, calcium, and other trace elements are also present in this fraction. In 2008, Maeba et al. [40] gave inositol orally to obese people for two weeks and observed improvement in waist circumference, high-sensitivity CRP, and fasting blood glucose level. He suggested plasmalogen was a key factor mediating the beneficial effect of inositol on metabolic syndrome. Myoinositol is a ring-shaped polyalcohol and element of the vitamin B complex. Taking large amounts of inositol (more than 10 g daily) could improve panic syndrome [41]. Phytic acid is phosphatized inositol and has a solid chelating and pH adjustment effect.

2.3.3 Healthy feeling and rice eating

Large-germ brown rice and pre-germinated brown rice contain functional ingredients to prevent dementia. Large germ rice and pre-germinated brown rice contain a high amount of GABA. GABA and γ-oryzanol are involved in the metabolism of hypothalamic catecholamines. γ-Oryzanol is known to have anti-stress effects, to palliate menopausal disorders and dysautonomia. The curative effect of Alzheimer's disease and amelioration in muscular fatigue is also present. GABA and γ-oryzanol have many activities, but the main hope is to improve cognitive function [42, 43].

The effect of ferulic acid mixed with Angelic archangelica extract on cognitive functions and behavioral and psychological symptoms of dementia has been examined by Kimura et al. [44], and many symptoms were shown to improve.

Pregnant women often become unstable in their mood. In an intervention study, pregnant women were randomized to take germinated rice or white rice for 14 days. They carried out a psychological test profile of mood states (POMS) before and after the study and employed salivary amylase as a stress marker.

Other components of rice bran, like steryl glucosides, were found to be effective for coping with stress. So, medical rice for mental health is at least defined to contain high GABA and γ-Oryzanol or ferulic acid.

Otsubo et al. [7] adopted the blend of super-hard genmai, wax-free genmai, and ordinary genmai at the rate of 4:4:2, adding 2.5% bran and 0.3% rice oil for the taste. This boiled rice is rich in dietary fiber, anthocyanin, and free ferulic acid and shows β-secretase inhibitory activity. A randomized study on 24 subjects for 12 weeks exhibited significant improvement in language memory by cognitive test.

3. Rice eating, gut microbiota, and short chain fatty acids

3.1 Rice eating and intestinal microbiota

Our study in Japan and community-wide analyses of the gut microbiota in Ireland showed that the intestinal microbiota varied by their dietary habits and living conditions [45, 46]. Intestinal microbiota change very rapidly in cases of fasting and Ayurveda (*Virechana* and *Basti*) [47, 48].

Numerous human studies, however, consistently demonstrated that gut microbiota could modulate host health [49, 50]. We focused on the people who had eaten brown rice in 2016 and 2017 (GENKI study I). We found that genmai eaters keep a healthy body weight (BMI 22.5) and have a good bowel movement. They preferred plant-based Japanese foods, avoiding meat and dairy products [26]. They disliked the oily and spicy taste and selected new, organic, with no additives, without genetically modified foodstuff, or domestic production. One hundred nine of them (male: 18; 50.1 ± 15.1 years old, female: 91, 55.8 ± 13.8 years old) agreed to examine the fecal bacteria [45].

Common bacterial profiles at phylum level were *Fermicutes* 44.3 *Bacteroides* 20.7 ± ±9.9%, 8.8%, *Actinobacteria* 8.3 ± 6.3%, *Proteobacteria* 1.7 ± 2.7%, and *Verrucobacteria* 1.2 ± 4.2% (max 39.4%). It represented only a minute fraction of gut microbiota, but they showed a characteristic pattern of rice eaters. The number of genera more than 1.0% was 15, more than 0.01%, and max more than 1.0% was 57 among 214 genera. The more than 80% prevalence of microbiota among participants was 31 (**Figure 1**). Common bacteria more than 1% profile were *Bacteroides* (12.7%), *Blautia* (8.3%), *Faecalibacterium* (7.9%), *Bifidobacterium* (6.3%), *Prevotella* (5.3%), *Eubacterium* (4.9%), *Ruminococcus* (3.8%), *Fusicatenibacter* (2.6%), *Collinsella* (1.9%), *Streptococcus* (2.4%), *Subdoligranulum* (2.1%), *Anaerostipes* (1.7%), *Ak*kermansia (1.2%), and *Roseburia* (1.7%). Individual variety was large, and we had to be careful about analyzing the network among the microbiota.

Correlation among these bacteria showed that *Firmicutes* negatively correlated with all other phyla. Although *Lachnoclocstridium Lactobacillus, Interstinibacter, Enterococcus, Mitsuokella, Tyzzerella*, and *Erysipelatoclostrium* in *Firmicutes* phylum, and *Alistipes* and *Odoribacter* in *Bacteroides* phylum showed a significant correlation. *Ruminococcus, Fusicaternibacter*, and *Anaerostipes* negatively correlated with the above genera. *Bifidobacterium* only showed a positive correlation with *Fusobacterium*.

Bacterial communication for growth should be further analyzed in the future [46–48].

3.2 Short chain fatty acids and gut environment (immunity)

Intestinal microbiota produces short chain fatty acids, such as acetate, propionate, butyrate, valeric, and caproate [49–52].

Genmai diets contain high levels of nondigestible dietary fiber, passing through the small intestine and leading to bacterial fermentation in the colon. The mildly acidic condition in the proximal colon fits butyrate production [50]. From the recent intestinal bacterial research, short chain fatty acid, especially butyrate-producing bacteria, is a focus of studies. Butyrate becomes an energy source for the colonic epithelial cells, maintains the gut barrier functions, and develops immune regulation and anti-inflammatory properties [52].

Butyrate was produced not only from dietary fiber but also from lactate, encompassing *A. caccae, A. butyraticus, A. hadrus, and E. hallii*. Furthermore, bacteria related to *Eubacterium hallii* and *Anaerostipes caccae* convert acetate and lactate into butyrate.

We had done the intervention study with a brown rice lunch five times/week as a business lunch for 12 weeks [51]. The results suggested that brown rice-eating induced stable innate immunity by short-chain fatty acids (SCFA), which stimulated the proliferation of regulatory T cells [52–54]. In that study, brown rice genmai omusubi (rice cake) was provided. We measured IL-6, CRP, and TNFα, as inflammatory markers for correlation analysis with microbiota changes.

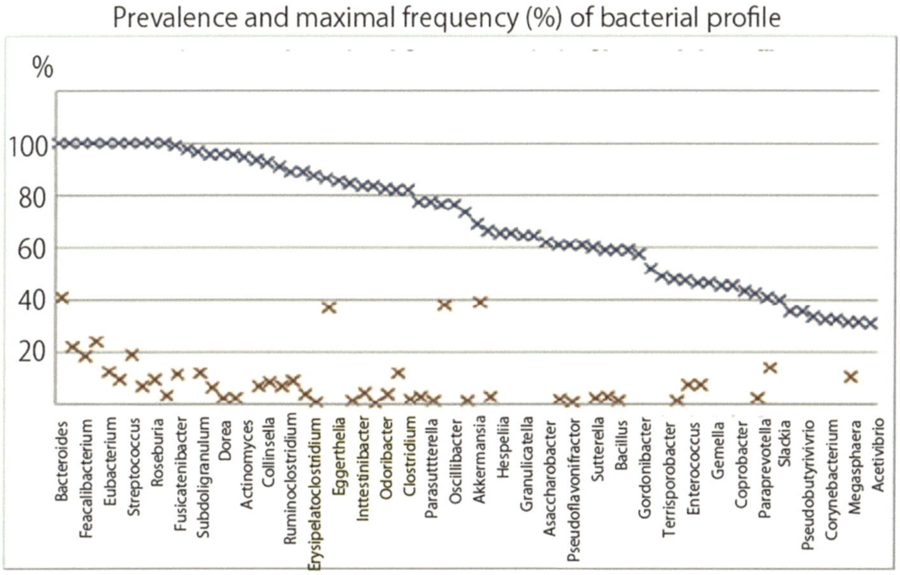

Figure 1.
Prevalence and maximal frequency (%) of bacterial profile among rice eaters. All people have Bacteroides, Faecalibacterium, Eubacterium, Streptococcus, Roseburia, Fusicatenibacter. More than one-third of rice eaters have the common 41 bacteria. Honda's [53] eleven microbiota for adaptive immune reaction are Parabacterium distasonis, Parabacterium gordonii, Alistipes senegalensis, Parabacteroides johnsonii, Parabacteroides xylaniphila, Bacteroides dorei, Bacteroides uniformis JCM5823, Eubacterium limosum, Rumonococcaceae bacterium cv2, Phascolarctobacterium faecium, and Fusobacterium ulcerans. Most people in our study have these bacteria. About 70% of people have Akkermancia [middle]. x shows max occupancy among people.

After three months of eating brown rice lunch, the body weight decreased in about half of the participants, and bowel movements and stool status improved significantly [51]. Brown (genmai) rice favored a gut microbiota with highly prevalent *Firmicutes* and a low prevalence of *Fusobacterium*. Significant microbiotic change was an increase in *Actinobacteria* and a decrease in *Proteobacteria. Blautia wexlerae, Collinsella aerofacience,* and *E. hallii* significantly increased at the species level.

Kenya Honda [53] found 11 rare, low-frequent human microbiome compositions suggesting potential biotherapeutics. Brown rice eaters have most of these, and the difference between brown and white rice eaters by microbiota profile was in high butyrate production. The upper tertial of genmai eaters tended to show low IL-6 and CRP, while TNFα was high. Butyrate binds the GRP109 receptor of the epithelial surface, which signal is transferred to the submucosal layer and stimulates the proliferation and maturation of regulatory T cells [54–56].

We recognized the importance of dendritic cells as the first playmaker because they were tissue-fixed cells and kept intimate communication with T cells [57]. We studied the maturation and distribution of dendritic cells in the human fetus. They first appeared in the thymus of a two-month-old human fetus and then spread to the peripheral lymphoid tissues with T lymphocytes. This movement was independent of the monocyte–macrophage lineage.

Dendritic cells and macrophages processed viral antigens to T4, and T8 lymphocytes, which secreted cytokines and chemokines to cause tissue inflammation. If the secreted cytokines were overshooting [58, 59], any suppressive mechanisms to stabilize immune reaction should be present to avoid severe progression.

4. Processed rice with Japanese agriculture standard (JAS)

4.1 Need of dietary therapy

Kaibara Ekiken (1630–1714) and Ishizuka Sagen's (1851–1909) dietary regimen is the traditional backbone of the Japanese well-being practice. The oriental worldview is more modest and emphasizes sustainability as much as possible. The traditional regimens influence oriental medicine, which promotes dietary therapy for chronic diseases. Integrative nutriology and medicine are grounded in this cultural background.

The longevity of the Japanese provides historical evidence of the above regimens. The Cabinet Office certifies foods for sick people and functional foods in Japan. The Ministry of Agriculture, Forestry, and Fisheries has decided on JAS (Japan Agriculture Standard) and is trying to expand it internationally.

A good example of the necessity of dietary therapy is chronic kidney disease (CKD).

The number of CKD patients is increasing worldwide [60]. The low protein diet was a traditional treatment to decrease proteinuria [61–64], and it was the only method until the 1960s' when invented hemodialysis. Since the 1980s, the development of antihypertensive, antidiabetic, diuretics, and recent SGL-2 inhibitors discarded the usefulness of low protein diet therapy. Even though, these therapies have not decreased the number of end-stage renal disease (ESRD) [65]. Rigorous multidrug treatment in the FROM-J and Doit3 studies in Japan did not show a significant reduction in a complication of CKD in diabetic patients [66, 67]. Recently, the negative spiral of the gut-kidney axis has been found behind CKD [68]. Uremic dysbiosis is associated with endotoxemia and chronic inflammation, disrupting the intestinal barrier and depletion of beneficial bacteria producing short-chain fatty acids. So, it is impossible to halt CKD progression without controlling uremic dysbiosis and leaky gut.

4.2 Low protein fermented genmai (Manufacturing Process Control JAS for LPFG)

We succeeded in deleting the rice protein from the brown rice bran by a particular combination of enzyme solution and *Lactobacillus plantaris* [9, 68, 69]. In addition to the remaining carbohydrate (energy source), low protein, low potassium, low phosphate character of low-protein rice, presence of dietary fiber, γ-oryzanol, and antioxidant activity were the characteristic biomarkers of brown rice.

The combination of four steps was the optimal process to produce LPFG [9, 68]. In *Lactobacillus* fermentation, the ability to produce lactic acid was high, and the pH was rapidly reduced in the early fermentation stage, resulting in an acidic environment. In the optimum pH range, protease activity increased, proteolysis was promoted, and a sharply decreased pH suppressed the growth of other bacteria.

This processed or fermented low protein genmai (LPFG) is approved for the product and process JAS (Japan Agriculture Standard) from the Ministry of Agriculture, Fisheries, and Forestry. These improved the negative spiral of the gut-kidney axis caused by uremic dysbiosis and leaky gut.

Dietary therapy for CKD patients must simultaneously control energy source intake and protein restriction. We asked the patients to replace their staple foods with an LPFG package without severely restricting protein from side dishes. A preliminary intervention study of 3 months of LPFG improved constipation by increasing *B. wexlerae*, *Bifidobacteria*, acetic acid, and decreasing harmful bacteria. As expected, the protein intake decreased from 60 g to 50 g daily. If 60 kg man eat 48 g protein (0.8 g/kg body weight), 10 g decrease becomes 38 g (0.62 g/kg). If a 50 kg woman takes 40 g protein, 10 g decreased intake becomes 30 g (0.6 g/kg). So, people can easily practice a low protein diet to decrease protein intake.

There are several RCTs about the low protein diet, but the results are still debatable [65, 69, 70] . The programmed protein intake often became over at the end of the study, and the difference between control became small. Or total energy intake often becomes insufficient by reducing the diet. So, we would say that almost all RCTs failed due to difficulty keeping the programmed amount of protein and energy source intake throughout the study period.

Individual difference is more significant if we consider the gut-kidney axis and other intrabody metabolic networks. Taste preferences and receptivity to dietary advice are also influenced by personality [71].

As RCTs could not be performed successfully, the basis of the guideline is unstable in CKD. We would start the LPFG intervention study on CKD patients through the pro- and post-comparison study. Diet is essential for patients with renal insufficiency, but compliance is not easy. However, a diet that only substitutes white rice for the LPFG package and has no strict limitations for side dishes is easy to maintain good adherence to protein control.

Comparing pre-and post-assessment is a more straightforward and practical method under the solution-oriented strategy. Diet therapy is the key to success through the patient's self-reliant will and is suitable for efforts involving patients. CKD's silent nature, with its unpredictable symptoms, is a significant barrier to motivating patients' behavioral changes and therapeutic decision-making by healthcare providers [72–74].

4.3 Celiac disease and low gluterin rice

Celiac disease is a long-term autoimmune disorder that primarily affects the small intestine [75–77]. Classic symptoms include gastrointestinal problems such as chronic diarrhea, abdominal distention, malabsorption, loss of appetite, and failure to grow in normal children [78]. Non-classic symptoms are common in children older than two years [79]. There may be mild or absent gastrointestinal symptoms involving any part of the body or no apparent symptoms. A reaction to gluten causes coeliac disease, a group of various proteins found in wheat. Upon exposure to gluten, several different autoantibodies are made, which affect several organs [80, 81]. In the small intestine, this causes an inflammatory reaction and may cause villous atrophy.

The only known effective treatment is a strict lifelong gluten-free diet, which improves symptoms and reduces the risk of complications in most people [78]. It is estimated that 80% of cases remain undiagnosed, usually because of minimal or absent gastrointestinal complaints and a lack of knowledge of symptoms and diagnostic criteria [80].

The term "gluten-free" is not mean "complete absence." A recent systematic review concluded that consuming less than 10 mg of gluten daily is unlikely to cause histological abnormalities.

Regulation of the "gluten-free" label differs. The European Union issued rules in 2009 limiting the use of "gluten-free" labels for food products to those with less than

20 mg/kg of gluten and "very low gluten" labels for those with less than 100 mg/kg [82, 83]. The USFDA issued regulations in 2013 limiting the use of "gluten-free" labels for food products to those with less than 20 ppm of gluten [83]. The international Codex Alimentarius standard allows for 20 ppm of gluten in so-called "gluten-free" foods [84].

In October 2020, the Ministry of Agriculture, Forestry, and Fisheries announced that the product's gluten content should be one ppm or less by identifying where gluten may be mixed in the rice flour manufacturing process and preventing gluten from being mixed. The "Manufacturing Process Control JAS for Non-Gluten Rice Flour (JAS0014)" standard started for managing the manufacturing process. Regarding the JAS, since June 2021, the Japan Agricultural Standards Certification Alliance, which is a registered certification body, has just started certification.

Promoting the spread of Japanese gluten-free rice flour would help many patients in Europe and the USA.

5. Conclusion

Genmai eaters in the macrobiotic groups are usually calm and peaceful. Many functional ingredients are related to keeping healthy and preventing diseases, such as obesity, diabetes, hypertension, and impaired recognition. In addition to the health benefit, a recent study on the microbiota-gut-brain relationship is also a target for future research [74]. Diets focused on certain rice types influenced the bacterial profile and production of short-chain fatty acids. These affect the innate immunity and control immune response after that [85].

CKD is increasing worldwide, and recently it happened in an aging society and the diabetic complication and end-stage glomerulonephritis. A low-protein diet is the most effective intervention for the prevention of CKD. LPFG could provide enough energy with low protein, low potassium, and low phosphate, and brown rice's health benefits by dietary fiber, γ-oryzanol, and antioxidant activity for CKD patients. LPFG could be available for patients with renal insufficiency at any stage, and it may be the first clear target of "medical rice." Celiac disease, common in Europe and USA, can be controlled by a gluten-free rice powder. The gluten-free powder with JAS certification can yield many foods that assure celiac disease prevention.

Although the shreds of evidence of dietary therapy for various diseases are still insufficient, the concept of "medical rice" could be widened in the future (**Table 2**).

Medical rice for chronic kidney disease (CKD) (protein<1/20),
Medical rice for wellbeing (enough nutrients and organic culture).
Medical rice for diabetes (glycemic index<55),
Medical rice for mental health (high GABA, γ-oryzanol, and/or ferulic acid).
Medical rice for cancer prevention (high antioxidant capacity).

Organic cultivation is necessary to avoid toxic substances from fertilizers and insecticides and it contributes to natural environment. Fair trade for farmers is also employed by the purchase of rice at more than 800 yen/kg.

Table 2.
Candidate of medical rice.

Abbreviations

GABA	γ-aminobutylic acid
CKD	chronic kidney disease
LPFG	low protein fermented genmai
BMI	Body Mass Index
EPA	eicosapentaenoic acid
DHA	docosahexaenoic acid
HDL	high-density lipoproteins
RCT	Randomized Clinical Trials
POMS	Profile of Mood States
JAS	Japan Agriculture Standard

Author details

Shaw Watanabe
Tokyo University of Agriculture, Medical Rice Association, Tokyo, Japan

*Address all correspondence to: shahfa@upmc.edu

References

[1] Vanavichit A. Facts about rice. In: Sontag J, editor. Rice Processing: The Comprehensive Guide to Global Technology and Innovative Products. Germany: Verlag GmbH & Co. KG; 2014. pp. 11-24

[2] FAOSTAT. Data and preliminary 2013 data for 5 major commodity aggregates. 2012. Available from: http://faostat.fao.org/site/339/default.aspx. [Accessed July 13, 2016]

[3] Watanabe S. Principles of Nutritional Science. Tokyo, Japan: Nanko-do Co. Ltd; 2011

[4] Ishida E. Genmai-Brown Rice for Better Health. New York: Japan Pub Inc; 1989

[5] Watanabe S, Takahashi M, Hashimoto H, Kikuchi K, Matsuo M, Otsubo K. Medical rice, discovery of a new food. In: Singh RB, editor. Functional Foods and Nutraceuticals in Metabolic and Non-communicable Diseases. London: Elsevier

[6] Standard Tables of Food Composition iJapan. 2017. Available from: https://www.mext.go.jp › afieldfile › 2017/11/2815

[7] Nakamura S, Ikeuchi T, Araki A, Kasuga K, Watanabe K, Hirayama M, et al. Possibility for prevention of type 2. Diabetes mellitus and dementia. Using three kinds of brown rice blends after high-pressure treatment. Food. 2022;**11**:818

[8] Takei N, Watanabe N, Nakajo M. Low-protein rice (LPR) product: Processing method and product safety. Advanced Food Technolology Nutrition Science. 2017;**3**(1):33-41. DOI: 10.17140/AFTNSOJ-3-142

[9] Watanabe S, Minakuchi S, Yamaguchi M, Uchiyama H, Haramoto T, Egawa K, et al. A new low-protein foodstuff from processed brown rice for chronic kidney disease. Acta Scientific Nutrition Health. 2021;**5**(8):1-10

[10] Morita N, Maeda T. Gamma-aminobutyric acid and functional materials in brown rice and pseudo-cereal during germination and their applications to food processing. In: Tsukuba Agriculture Research Gallery, editor. Rice Studies, Present and Future. Tokyo, Japan: Sankyo Pub Co. Ltd; 2012. pp. 150-161

[11] Watanabe S, Kikuchi K, Mizuno S, Mizuno K. *Aspergillus oryzae* fermented brown rice and bran supplement "FBRA" for health and Cancer prevention. Acta Scientific Nutritional Health. 2021;**5**(12):20-32

[12] Kosaka H, Nemoto H, Motojima A. Antitumorigenic effect and antioxidative activity of the brown rice fermented by aspergillus oryzae (FBRA). In: Tsukuba Agriculture Research Gallery, editor. Rice Studies, Present and Future. Tokyo, Japan: Sankyo Pub Co. Ltd; 2012. p. 266

[13] Watanabe S, Hirakawa A, Nishijima C, Ohtsubo K, Nakamura K, Beppu S, et al. Food as medicine: The new concept of "medical rice". Advanced Food Technology Nutrition Science Open Journal. 2016;**2**(2):38-50. DOI: 10.17140/AFTNSOJ-2-129

[14] Hirakawa A, Watanabe S, Melby M. Comprehensive food labeling for obesity control. Advanced Obesity Weight Management. 2016;**4**(3):88. DOI: 10.15406/aowmc.2016.04.00088

[15] Watanabe S. Rice function for disease prevention and establishment of medical rice association. Diabetes Research Open Journal. 2019;**5**(1):e1-e3. DOI: 10.17140/DROJ-5-e013

[16] Standard Tables of Food Composition in Japan, 8th Edition. 2020. Available from: https://www.mext.go.jp › syokuhinseibun › mext_01110

[17] Sugano M. Functional characteristics of rice bran oil. Journal of Japan Society of Food Engineering. 2002;**22**:26-30

[18] Most MM, Tulley R, Morales S, Lefevre M. Rice bran oil, not fiber, lowers cholesterol in humans. American Journal of Clinical Nutrition. 2005;**81**(1):64-68

[19] Nakagami T, Gonda H, Hashimoto H, Tsuno T. LC/MS analysis of cerebrosides in rice bran. In: Tsukuba AgricultureResearch Gallery, editor. Rice Studies, Present and Future. Tokyo, Japan: Sankyo Pub Co. Ltd; 2012. pp. 225-232

[20] Ramazani E, Akaberi M, Emami SA, Tayarani-Najaran Z. Biological and pharmacological effects of gamma-oryzanol: An updated review of the molecular mechanisms. Current Pharmaceutical Design. 2021;**27**(19):2299-2316

[21] Somsuvra B, Shital J. Panchal gamma-oryzanol – A multi-purpose steryl ferulate. Current Nutrition & Food Science. 2011;**7**(1):10-20

[22] Kokumai T, Ito J, Kobayashi E, Shimizu N, Hashimoto H, Eitsuka T, et al. Comparison of blood profiles of γ-oryzanol and ferulic acid in rats after Oral intake of γ-oryzanol. Nutrients. 2019;**11**(5):1174. DOI: 10.3390/nu11051174

[23] Watanabe S, Mizuno S, Hirakawa A. Effects of brown rice on obesity: GENKI Study I (Cross Sectional Epidemiological Study). Journal of Obesity Chronic Diseases. 2019;**2**(1):12

[24] Kushi M, Jack A. The Macrobiotic Path to Total Health. New York, USA: Ballantine Books; 2004

[25] Kushi LH, Byers T, Doyle C, et al. American Cancer Society guidelines on nutrition and physical activity for cancer prevention: Reducing the risk of cancer with healthy food choices and physical activity. CA: A Cancer Journal for Clinicians. 2006;**56**:254-281. DOI: 10.3322/canjclin.56.5.254

[26] Takahashi M, Watanabe S, Mizuno S, Hirakawa A. Dietary and life habits of obesity and brown rice eaters among Genmai Evidence for Nutritional Kenko Innovation (GENKI) Study I and II. Acta Scientific Nutritional Health. 2020;**4**(8):21-32

[27] Tahira T. Prevention of dementia by improving lifestyle. Clinical Function Nutrition. 2012;**4**(6):296-301

[28] Sakomoto S, Hayashi T, Hayashi K, et al. Pre-germinated brown rice could enhance maternal mental health and immunity during lactation. European Journal of Nutrition. 2007;**46**:391-396. DOI: 10.1007/s00394-007-0678-3

[29] Matsushita K, Araki E, Kawase S, Sunohara Y, Iida S. *Haiibuki*, a rice variety with giant embryo, accumulates high amount of GABA. In: Tsukuba Agriculture Research Gallery, editor. Rice Studies, Present and Future. Tokyo, Japan: Sankyo Pub Co.Ltd; 2012. p. 314

[30] Shimabukuro M, Masuzaki H, Kinjo R, Yamakawa K, Tanaka H,

Kozuka C, et al. Effects of brown rice diet on visceral obesity and endothelial function: The BRAVO study. British Journal of Nutrition. 2013;**111**:310-320. DOI: 10.1017/S0007114513002432

[31] Smit HJ, Kemsley EK, Tapp HS, Henry CJK. Does prolonged chewing reduce food intake? Fletcher revisited. Appetite. 2011;**57**:295-298

[32] Tian S, Nakamura K, Cui T, Kayahara HJ. High-performance liquid chromatographic determination of phenolic compounds in rice. Journal of Chromatography. A. 2005;**1063**(1-2):121-128

[33] Tian S, Nakamura K, Kayahara H. Analysis of phenolic compounds in white rice, brown rice, and germinated brown rice. Journal of Agricultural and Food Chemistry. 2004;**52**:4808-4813. DOI: 10.1021/jf049446f

[34] Prior RL, Hoang H, Gu L, et al. Assays for hydrophilic and lipophilic antioxidant capacity (oxygen radical absorbance capacity (ORAC(F.L.))) of plasma and other biological and food samples. Journal of Agricultural and Food Chemistry. 2003;**51**:3273-3279. DOI: 10.1021/jf0262256

[35] Prior RL, Wu X, Schaich K. Standardized methods for the determination of antioxidant capacity and phenolics in foods. DOI: 10.1021/jf0502698

[36] Tatsumi Y, Ishihara J, Morimoto M, Ohno Y, Watanabe S. For the JPHC FFQ validation study group. Seasonal differences in total antioxidant capacity intake from foods consumed by a Japanese population. European Journal of Clinical Nutrition. 2014;**68**(7):799-803. DOI: 10.1038/ejcn.2014.65

[37] Miyazawa T, Sibata A, Nakagawa K, Phumon S. Cancer prevention by rice bran tocotrienol. In: Tsukuba Agriculture Research Gallery, editor. Rice Studies, Present and Future. Tokyo, Japan: Sankyo Pub Co. Ltd; 2012. pp. 217-224

[38] Grases F, Costa-bauza A, Prieto RM. Phytate as modulator of calcification processes in humans. In: Tsukuba Agriculture Research Gallery, editor. Rice Studies, Present and Future. Tokyo, Japan: Sankyo Pub Co. Ltd; 2012. pp. 167-177

[39] Vucenik I. IP6 and inositol in cancer prevention and therapy. In: Tsukuba Agriculture Research Gallery, editor. Rice Studies, Present and Future. Tokyo, Japan: Sankyo Pub Co. Ltd; 2012. pp. 196-208

[40] Maeba R, Hara H. Myo-inositol and plasmalogens. In: Tsukuba Agriculture Research Gallery, editor. Rice Studies, Present and Future. Tokyo, Japan: Sankyo Pub Co. Ltd; 2012. pp. 178-188

[41] Benjamin J, Levine J, Fux M, Aviv A, Levy D, Belmaker RH. Double-blind, placebo-controlled, crossover trial of inositol treatment for panic disorder. The American Journal of Psychiatry. 1995;**152**(7):1084-1086. DOI: 10.1176/ajp.152.7.1084

[42] Tungtrakul P. Innovative rice products in Thailand. In: Proceedings of the East Asia Conference on Standardization of Rice Function. Kyoto, Japan; 2013. pp. 61-62

[43] Muran V, Spiegelman D, Sadha V, et al. Effect of brown rice, white rice, and brown rice with legumes on blood glucose and insulin responses in overweight Asian Indians: A randomized control trial. Diabetes Technology & Therapeutics. 2014;**16**:317-325. DOI: 10.1089/dia.2013.0259

[44] Kimura T, Hayashida H, Murata M, Takamatsu J. Effect of ferulic

acid and angelica archangelica extract on behavioral and psychological symptoms of dementia in frontotemporal lobar degeneration and dementia with Lewy bodies. Geriatric International. 2011;**11**:309

[45] Hirakawa A, Aoe S, Watanabe S, Hisada T, Mochizuki J, Mizuno S, et al. The nested study on the intestinal microbiota in GENKI study with special reference to the effect of brown rice eating. Journal of Obesity Chronic Diseases. 2019;**3**(1):l1-l13

[46] Claesson MJ, Jeffery IB, Conde S, Power SE, O'Connor EM, Cusack S, et al. Gut microbiota composition correlates with diet and health in the elderly. Nature. 2012;**488**(7410):178-184. DOI: 10.1038/nature11319

[47] Watanabe S, Hirakawa A, Utada I, Aoe S, Moriyama S, Hitomi S, et al. Ketone body production and excretion during wellness fasting. Diabetes Research Open Journal. 2017;**3**(1):1-8. DOI: 10.17140/DROJ-3-129

[48] Watanabe S, Oikawa S, Mibayashi H. Ayurveda treatment (*Virechana and Basti*) and changes of intestinal microbiota at phyla and species level. Prensa Médica Argentina. 2021;**107**(3):319. DOI: org/10.47275/0032-745X-319

[49] Velasquez-Manoff M. The peacekeepers. Nature. 2015;**518**:S3-S5

[50] Louis P, Flint HJ. Diversity, metabolism and microbial ecology of butyrate-producing bacteria from the human large intestine. FEMS Microbiology Letters. 2009;**294**(1):1-8. DOI: 10.1111/j.1574-6968.2009.01514.x

[51] Kikuchi K, Watanabe S, Matsuo M, Ezaki T, Mizuno S, Hisada T, et al. Changes in microbiota and short-chain fatty acids following 3-month pilot intervention study feeding brown rice ball (Omusubi) to healthy volunteers. La Prensa Medica Argentina. 2020;**107**(1):1-11

[52] Furusawa Y, Obata Y, Fukuda S, Endo T, Nakato G, Takahashi D, et al. Commensal microbe-derived butyrate induces the differentiation of colonic regulatory T cells. Nature. 2013;**504**(7480):446-450

[53] Honda K, Littman DR. The microbiota in adaptive immune homeostasis and disease. Nature. 2016;**535**(7610):75-84. DOI: 10.1038/nature18848

[54] Smith PM, Howitt MP, Panikov N, Michaud M, Gallini CA, et al. The microbial metabolites, short-chain fatty acids, regulate colonic Treg cell homeostasis. Science. 2013;**341**:69-57

[55] Bachem A, Makhlouf C, Binger KJ, de Souza DP, Tull D, Hochheiser K, et al. Microbiota-derived short-chain fatty acids promote the memory potential of antigen-activated CD8(+) T cells. Immunity. 2019;**51**:285-297

[56] Kimura I, Inoue D, Maeda T, Hara T, Ichimura A, Miyauchi S, et al. Short-chain fatty acids and ketones directly regulate sympathetic nervous system via G protein-coupled receptor 41 (GPR41). Proceedings of the National Academy of Sciences of the United States of America. 2011;**108**:8030-8035. DOI: 10.1073/pnas.1016088108

[57] Watanabe S, Nakajima T, Shimosato Y, Shimamura K, Sakuma H. T-zone histiocytes with S100 protein. Pathology International. 1983;**33**(1):15-22

[58] Henderson LA, Canna SW, Schulert GS, Volpi S, Lee PY, Kernan KF, et al. On the alert for cytokine storm:

Immunopathology in COVID-19. Arthritis & Rhematology. 2020;**72**:1059-1063. DOI: 10.1002/art.41285

[59] Mehta P, McAuley DF, Brown M, Sanches E, Tattersall RS, Manson JJ. COVID-19: Consider cytokine storm syndromes and immunosuppression. Lancet. 2020;**395**:1033-1034. DOI: 10.1016/s140-6736(20)30628-0

[60] GDB Chronic Kidney Disease Collaboration. Global, regional, and national burden of chronic 249 kidney disease, 1990-2017: A systematic analysis for the global burden of disease study 250 2017. Lancet. 2020;**395**:709-733

[61] Kopple JD, Massry SG. Nutritional Management of Renal Disease. Baltimore, Maryland, USA: Williams & Wilkins; 1997

[62] Kopple JE, Levey A, Greene T, et al. Effect of dietary protein restriction on nutritional status in the modification of diet in renal disease study. Kidney International. 1997;**52**:778-791. DOI: 10.1038/ki.1997.395

[63] Watanabe S. Recipe of Low Protein Diet to Keep Kidney Function. Tokyo, Japan: Shufu-no-tomo-sha; 2010

[64] Watanabe S. Low-protein diet for the prevention of renal failure. Proceedings of the Japan Academy. 2017;**93**:1-9

[65] Levy AS, Greene T, Beck GJ, Caggiula AW, Kusek JW, Hunsicker LG, et al. Dietary protein restriction and the progression of chronic renal disease: What have all of the results of the MDRD study shown? Journal of American Soiety Nephrology. 1999;**10**:2426-2439

[66] Saito C, Yamagata K. Multidisciplinary care for CKD in the FROM-J study. Nihon Jinzo Gakkai Shi. 2015;**57**(5):811-818

[67] Ueki K, Sasako T, Okazaki Y, Kato M, Okahata S, Katsuyama H, et al. Effect of an intensified multifactorial intervention on cardiovascular outcomes and mortality in type 2 diabetes (J-DOIT3): An open-label, randomised controlled trial. The Lancet Diabetes and Endocrinology. 2017;**5**:951-964. DOI: 10.1016/S2213-8587(17)30327-3

[68] Watanabe S. Dietary therapy with low protein genmai (brown rice) to improve the gut-kidney axis and reduce CKD progression. APJCN. 2022;**3**(3):(in press)

[69] Watanabe S, Ohtsubo K. Low-protein diet: History and use of processed low-protein rice for the treatment of chronic kidney disease. Food. 2021;**10**:2255. DOI: 10.3390/foods10102255

[70] Lu L, Huang YF, Wang M-Q, Chen D-X, Wan H, Wei L-B, et al. Dietary fiber intake is associated with chronic kidney disease (CKD) progression and cardiovascular risk, but not protein nutritional status, in adults with CKD. Asia Pacific Journal of Clinical Nutrition. 2017;**26**:598-605

[71] Ohmori Y, Suzuki N, Morita A, Aiba A, Miyachi M, Watanabe S for SCOP Group. Association of personality (NEO-Five Factor Inventory) with eating behaviors and physical activity levels in obese subjects in the Saku Control Obesity Program (SCOP). Anti-Aging Medicine. 2007;**4**(2):43-50

[72] Vaziri ND, Liu S-M, Lau WL, Khazaeli M, Nazertehrani S, Farzaneh SH, et al. High amylose resistant starch diet ameliorates oxidative stress, inflammation, and progression of chronic kidney disease. PLoS One. 2014;**9**:e114881

[73] Buchanan S, Combet E, Stenvinke P, Shiels PG. Klotho, aging, and the failing

kidney. Frontiers in Endocrinology. 2020;**11**:355

[74] Masuzaki H, Uema T, Millman JF, Okamoto S. The power of gut-brain interaction as a promising target for healthy longevity. Glycative Stress Research. 2022;**9**(1):1-6. DOI: 10.24659/gsr.9.1_1 online : March 31, 2022

[75] Marsh MN. Gluten, major histocompatibility complex, and the small intestine. A molecular and immunobiologic approach to the spectrum of gluten sensitivity (celiac sprue). Gastroenterology. 1992;**102**(1):330-354. DOI: 10.1016/0016-5085(92)91819

[76] Shan L, Qiao SW, Arentz-Hansen H, Molberg Ø, Gray GM, Sollid LM, et al. Identification and analysis of multivalent proteolytically resistant peptides from gluten: Implications for celiac sprue. Journal of Proteome Research. 2005;**4**(5):1732-1741. DOI: 10.1021/pr050173t

[77] Koning F, Schuppan D, Cerf-Bensussan N, Sollid LM. Pathomechanisms in celiac disease. Best Practice & Research. Clinical Gastroenterology. 2005;**19**(3):373-387. DOI: 10.1016/j.bpg.2005.02.003

[78] van der Windt DA, Jellema P, Mulder CJ, Kneepkens CM, van der Horst HE. Diagnostic testing for celiac disease among patients with abdominal symptoms: A systematic review. Journal of the American Medical Association. 2010;**303**(17):1738-1746. DOI: 10.1001/jama.2010.549

[79] Wong RC, Steele RH, Reeves GE, Wilson RJ, Pink A, Adelstein S. Antibody and genetic testing in coeliac disease. Pathology. 2003;**35**(4):285-304. DOI: 10.1080/00313020307527

[80] Castillo NE, Theethira TG, Leffler DA. The present and the future in the diagnosis and management of celiac disease. Nutrition in Clinical Practice (Review). 2015;**3**(1):3-11. DOI: 10.1093/gastro/gou065

[81] García-Manzanares A, Lucendo AJ. Nutritional and dietary aspects of celiac disease. Nutrition in Clinical Practice (Review). 2011;**26**(2):163-173. DOI: 10.1177/0884533611399773

[82] Lee AR, Ng DL, Zivin J, Green PH. Economic burden of a gluten-free diet. Journal of Human Nutrition and Dietetics. 2007;**20**(5):423-430

[83] van Berge-Henegouwen GP, Mulder CJ. Pioneer in the gluten free diet Willem-Karel Dicke 1905-1962, over 50 years of gluten free diet. Gut. 1993;**34**(11):1473-1475. DOI: 10.1136/gut.34.11.1473

[84] Coeliac disease. Available from: https://en.wikipedia.org/wiki/Coeliac_disease

[85] Watanabe S, Iinuma K. Rice Consumption enhancing innate immunity with a reduced risk of COVID-19 infection and mortality. Acta Scientific Nutritional Health. 2022;**6**(4):80-90